Statistical Data Modeling and Machine Learning with Applications

Statistical Data Modeling and Machine Learning with Applications

Editor

Snezhana Gocheva-Ilieva

MDPI • Basel • Beijing • Wuhan • Barcelona • Belgrade • Manchester • Tokyo • Cluj • Tianjin

Editor
Snezhana Gocheva-Ilieva
University of Plovdiv Paisii Hilendarski
Bulgaria

Editorial Office
MDPI
St. Alban-Anlage 66
4052 Basel, Switzerland

This is a reprint of articles from the Special Issue published online in the open access journal *Mathematics* (ISSN 2227-7390) (available at: https://www.mdpi.com/journal/mathematics/special_issues/Statistical_Data_Modeling_Machine_Learning_Applications).

For citation purposes, cite each article independently as indicated on the article page online and as indicated below:

LastName, A.A.; LastName, B.B.; LastName, C.C. Article Title. *Journal Name* **Year**, *Volume Number*, Page Range.

ISBN 978-3-0365-2692-8 (Hbk)
ISBN 978-3-0365-2693-5 (PDF)

© 2021 by the authors. Articles in this book are Open Access and distributed under the Creative Commons Attribution (CC BY) license, which allows users to download, copy and build upon published articles, as long as the author and publisher are properly credited, which ensures maximum dissemination and a wider impact of our publications.

The book as a whole is distributed by MDPI under the terms and conditions of the Creative Commons license CC BY-NC-ND.

Contents

Snezhana Gocheva-Ilieva
Special Issue "Statistical Data Modeling and Machine Learning with Applications"
Reprinted from: *Mathematics* **2021**, *9*, 2997, doi:10.3390/math9232997 1

Snezhana Gocheva-Ilieva, Hristina Kulina and Atanas Ivanov
Assessment of Students' Achievements and Competencies in Mathematics Using CART and CART Ensembles and Bagging with Combined Model Improvement by MARS
Reprinted from: *Mathematics* **2021**, *9*, 62, doi:10.3390/math9010062 5

Trung Duc Tran, Vinh Ngoc Tran and Jongho Kim
Improving the Accuracy of Dam Inflow Predictions Using a Long Short-Term Memory Network Coupled with Wavelet Transform and Predictor Selection
Reprinted from: *Mathematics* **2021**, *9*, 551, doi:10.3390/math9050551 23

Tat'y Mwata-Velu, Jose Ruiz-Pinales, Horacio Rostro-Gonzalez, Mario Alberto Ibarra-Manzano, Jorge Mario Cruz-Duarte and Juan Gabriel Avina-Cervantes
Motor Imagery Classification Based on a Recurrent-Convolutional Architecture to Control a Hexapod Robot
Reprinted from: *Mathematics* **2021**, *9*, 606, doi:10.3390/math9060606 45

Jose Torres-Pruñonosa, Pablo García-Estévez and Camilo Prado-Román
Artificial Neural Network, Quantile and Semi-Log Regression Modelling of Mass Appraisal in Housing
Reprinted from: *Mathematics* **2021**, *9*, 783, doi:10.3390/math9070783 61

Jesus Cerquides, Mehmet Oğuz Mülâyim, Jerónimo Hernández-González, Amudha Ravi Shankar and Jose Luis Fernandez-Marquez
A Conceptual Probabilistic Framework for Annotation Aggregation of Citizen Science Data
Reprinted from: *Mathematics* **2021**, *9*, 875, doi:10.3390/math9080875 77

Aurea Grané and Alpha A. Sow-Barry
Visualizing Profiles of Large Datasets of Weighted and Mixed Data
Reprinted from: *Mathematics* **2021**, *9*, 891, doi:10.3390/math9080891 93

Jianli Shao, Xin Liu and Wenqing He
Kernel Based Data-Adaptive Support Vector Machines for Multi-Class Classification
Reprinted from: *Mathematics* **2021**, *9*, 936, doi:10.3390/math9090936 113

Jingcheng Zhou, Wei Wei, Ruizhi Zhang and Zhiming Zheng
Damped Newton Stochastic Gradient Descent Method for Neural Networks Training
Reprinted from: *Mathematics* **2021**, *9*, 1533, doi:10.3390/math9131533 129

Ala'a El-Nabawy, Nahla A. Belal and Nashwa El-Bendary
A Cascade Deep Forest Model for Breast Cancer Subtype Classification Using Multi-Omics Data
Reprinted from: *Mathematics* **2021**, *9*, 1574, doi:10.3390/math9131574 141

Hui Chen, Kunpeng Xu, Lifei Chen and Qingshan Jiang
Self-Expressive Kernel Subspace Clustering Algorithm for Categorical Data with Embedded Feature Selection
Reprinted from: *Mathematics* **2021**, *9*, 1680, doi:10.3390/math9141680 155

Editorial

Special Issue "Statistical Data Modeling and Machine Learning with Applications"

Snezhana Gocheva-Ilieva

Department of Mathematical Analysis, University of Plovdiv Paisii Hilendarski, 4000 Plovdiv, Bulgaria; snow@uni-plovdiv.bg

Citation: Gocheva-Ilieva, S. Special Issue "Statistical Data Modeling and Machine Learning with Applications". *Mathematics* **2021**, *9*, 2997. https://doi.org/10.3390/math9232997

Received: 16 November 2021
Accepted: 18 November 2021
Published: 23 November 2021

Publisher's Note: MDPI stays neutral with regard to jurisdictional claims in published maps and institutional affiliations.

Copyright: © 2021 by the author. Licensee MDPI, Basel, Switzerland. This article is an open access article distributed under the terms and conditions of the Creative Commons Attribution (CC BY) license (https://creativecommons.org/licenses/by/4.0/).

Give Us Data to Predict Your Future!

The modeling and processing of empirical data is one of the main subjects and goals of statistics. Nowadays, with the development of computer science, the extraction of useful and often hidden information and patterns from data sets of different volumes and complex data sets in warehouses has been added to these goals. New and powerful statistical techniques with machine learning (ML) and data mining paradigms have been developed. To one degree or another, all of these techniques and algorithms originate from a rigorous mathematical basis, including probability theory and mathematical statistics, operational research, mathematical analysis, numerical methods, etc. Popular ML methods, such as artificial neural networks (ANN), support vector machines (SVM), decision trees, random forest (RF), among others, have generated models that can be considered as straightforward applications of optimization theory and statistical estimation. The wide arsenal of classical statistical approaches combined with powerful ML techniques allow many challenging and practical problems to be solved.

This Special Issue belongs to the section "Mathematics and Computer Science". Its aim is to establish a brief collection of carefully selected papers presenting new and original methods, data analyses, case studies, comparative studies, and other research on the topic of statistical data modeling and ML as well as their applications. Particular attention is given, but is not limited, to theories and applications in diverse areas such as computer science, medicine, engineering, banking, education, sociology, economics, among others.

This Special Issue begins with a contribution to computer science and educational data mining (EDM) by Gocheva et al. [1]. A new framework based on three ML regression methods for predicting student achievement in mathematics through the statistical processing of empirical data is proposed. In the first stage, classification and regression trees (CART) as well as CART ensembles and bagging models are built and evaluated. The importance of predictors for student success is determined. In the next stage, the predicted values of the best models from the first stage are combined through stacking with the multivariate adaptive regression splines (MARS) method. It is shown that the combined MARS models are superior to the single models for all of the statistical indicators.

An engineering application to predict dam inflow levels by means of time series modeling using a long short-term memory (LSTM) network is developed in the work of Tran et al. [2]. A robust statistical criterion called the "correlation threshold" for partial autocorrelation and cross-correlation functions is introduced to the appropriate input predictors and the number of their time lag variables. A wavelet transformation and a hyper-parameter optimization determined by K-fold cross-validation were also applied to improve the overall performance of the LSTM. The resulting streamflow predictions are shown to be more accurate than those of ANNs, recurrent NNs, support vector regression, and multilayer perceptron (MLP).

The application of data modeling in medicine is presented in the work of Mwata-Velu et al. [3]. The problem of information processing in the interactions between brain signals and controllable machines, which requires instantaneous brain data decoding, is considered. The authors have developed a real-time embedded brain–computer interface (BCI) system for signal recognition, and this system is applied in the locomotion of a hexapod robot. A hybrid convolutional neural network (CNN)–LSTM model is implemented to process and classify motor imagery electroencephalogram signals that have been captured by specialized sensors. By using stratified 10-fold cross-validation, the average model accuracy is determined to be about 85%.

The next article in this Special Issue models the hedonistic prices of the housing market in Catalonia. The authors Torres-Pruñonosa et al. [4] analyze a large amount of data provided by two banks. Regression models with ANN, quantile regression (QR), and semi-logarithmic regression (SLR) were obtained and studied. A comparison of the results showed that QR models are less accurate than other models, and therefore, the QR method cannot be recommended. The ANN and SLR models demonstrated similar statistics, with SLR showing the best performance in the case of smaller data volumes. The study offers recommendations to Spanish banks, encourage the use of ANN or SLR for real estate appraisal.

Interesting results have been reported in [5] by Cerquides and his co-authors. This study contributes to the crowdsourcing paradigm based on "the wisdom of crowd". In particular, this article deals with methods for assessing the reliability and accuracy of public data that have been collected and analyzed by volunteers—members of different citizen science communities. An abstract probabilistic consensus model is proposed, which summarizes other label aggregation models. Practical evidence supporting this approach is demonstrated by assessing the accuracy of crowdsourced graphical data on the earthquake in Albania during 2019, difficult to process statistically by means of classical and ML statistical techniques.

The work of Grané and Sow-Barry [6] explores a hybrid approach that consists of two classical multivariate techniques—multidimensional scaling (MDS) and the k-prototype clustering algorithm—to visualize the profiles of individuals from weighted and mixed big data. Initially, a small random working sample (up to 10%) was extracted from the whole dataset. This sample was clustered using the k-prototype clustering algorithm with Gower's distance, the individuals were labeled trough weighted MDS, and the MDS configuration was determined. Then, the rest of the individuals were projected onto the MDS configuration by the Gower's interpolation. The algorithm was implemented to classify and visualize large survey data depicting the health and socioeconomic living conditions of European citizens.

The Special Issue continues with the work of Shao et al. [7], who focuses on the multi-category classification problem. A framework and ML algorithm based on data-adaptive kernel SVM for binary imbalanced data have been proposed, including a new method for constructing the data-dependent kernel for a multi-class setting. The improved performance of the method is mainly due to the more robust decay rate and flexibility of the constructed kernel functions. This ensures the optimal adaptation of the data-dependent kernel to the studied data. Tests were performed with different datasets, both artificial and real. An application of the new method is conducted in the medical field for the classification of a multi-class cancer imaging dataset. The simulations illustrate that the model results are significantly better compared to six standard classifiers for all of the statistical indicators.

In paper [8], Zhou et al. develop two new second-order optimization methods, namely the damped Newton stochastic gradient descent (DN-SGD) method and the stochastic gradient descent-damped Newton (SGD-DN) method, which were designed to train deep neural networks (DNNs). A mathematical expression for calculating the Hessian matrices of the last layer parameters and the penultimate layer of the loss function is derived, and it is proven that the Hessian matrices are positive and semidefinite. Furthermore, the DN-SGD and SGD-DN algorithms are implemented to iterate the parameters of the last layer using a variational damped Newton method. In this way, faster algorithm convergence is achieved, and the cost of the calculations is reduced. Numerical experiments have been performed, which demonstrate the higher efficiency of the proposed methods for solving various regression and classification problems with data from the housing market, finance, and other areas.

A system for classifying breast cancer subtypes using the cascade deep forest method is presented in the work of El-Nabawy et al. [9]. The cascade deep forest model incorporates both DNNs and ensemble models such as RF. The model learns the class distribution features by assembling decision tree-based forests while supervising different types of input features. The classification is employed on various omics METABRIC sub-datasets that contain preprocessed and integrated data profiles of clinical datasets, gene expression datasets, and two types of datasets obtained through statistical feature engineering for different classes of dataset configurations. A model accuracy of up to 83% for 5 subtypes and of up to 78% for 10 subtypes was reported. It is shown that with the developed system, the time needed to obtain the results is shorted compared to the previous results in the field.

The Special Issue concludes with an article by Chen et al. [10], which proposes a new algorithm for the non-linear clustering of categorical data. The algorithm includes a self-expressive kernel density estimation scheme and a probability-based non-linear feature-weighted similarity measure. A non-linear optimization method in kernel subspace is implemented in the developed self-expressive kernel subspace clustering algorithm with embedded feature selection. Experiments were performed that highlighted the better performance of the algorithm compared to classical clustering algorithms used for categorical data with non-linear relationships.

In conclusion, the resulting palette of methods, algorithms, and applications for statistical modeling and ML presented in this Special Issue is expected to contribute to the further development of research in this area. We also believe that the new knowledge acquired here as well as the applied results are attractive and useful for young scientists, doctoral students, and researchers from various scientific specialties.

Funding: This research received no external funding.

Acknowledgments: The research activity of the Guest Editor of this Special Issue has been conducted in the framework and has been partially supported by the MES (Grant No. D01-387/18.12.2020) for NCDSC, part of the Bulgarian National Roadmap on RIs.

Conflicts of Interest: The author declares no conflict of interest.

References

1. Gocheva-Ilieva, S.; Kulina, H.; Ivanov, A. Assessment of Students' Achievements and Competencies in Mathematics Using CART and CART Ensembles and Bagging with Combined Model Improvement by MARS. *Mathematics* **2021**, *9*, 62. [CrossRef]
2. Tran, T.D.; Tran, V.N.; Kim, J. Improving the Accuracy of Dam Inflow Predictions Using a Long Short-Term Memory Network Coupled with Wavelet Transform and Predictor Selection. *Mathematics* **2021**, *9*, 551. [CrossRef]
3. Mwata-Velu, T.; Ruiz-Pinales, J.; Rostro-Gonzalez, H.; Ibarra-Manzano, M.A.; Cruz-Duarte, J.M.; Avina-Cervantes, J.G. Motor Imagery Classification Based on a Recurrent-Convolutional Architecture to Control a Hexapod Robot. *Mathematics* **2021**, *9*, 606. [CrossRef]
4. Torres-Pruñonosa, J.; García-Estévez, P.; Prado-Román, C. Artificial Neural Network, Quantile and Semi-Log Regression Modelling of Mass Appraisal in Housing. *Mathematics* **2021**, *9*, 783. [CrossRef]
5. Cerquides, J.; Mülâyim, M.O.; Hernández-González, J.; Shankar, A.R.; Fernandez-Marquez, J.L. A Conceptual Probabilistic Framework for Annotation Aggregation of Citizen Science Data. *Mathematics* **2021**, *9*, 875. [CrossRef]
6. Grané, A.; Sow-Barry, A.A. Visualizing Profiles of Large Datasets of Weighted and Mixed Data. *Mathematics* **2021**, *9*, 891. [CrossRef]
7. Shao, J.; Liu, X.; He, W. Kernel Based Data-Adaptive Support Vector Machines for Multi-Class Classification. *Mathematics* **2021**, *9*, 936. [CrossRef]
8. Zhou, J.; Wei, W.; Zhang, R.; Zheng, Z. Damped Newton Stochastic Gradient Descent Method for Neural Networks Training. *Mathematics* **2021**, *9*, 1533. [CrossRef]
9. El-Nabawy, A.; Belal, N.A.; El-Bendary, N. A Cascade Deep Forest Model for Breast Cancer Subtype Classification Using Multi-Omics Data. *Mathematics* **2021**, *9*, 1574. [CrossRef]
10. Chen, H.; Xu, K.; Chen, L.; Jiang, Q. Self-Expressive Kernel Subspace Clustering Algorithm for Categorical Data with Embedded Feature Selection. *Mathematics* **2021**, *9*, 1680. [CrossRef]

Article

Assessment of Students' Achievements and Competencies in Mathematics Using CART and CART Ensembles and Bagging with Combined Model Improvement by MARS

Snezhana Gocheva-Ilieva *, Hristina Kulina and Atanas Ivanov

Department of Mathematical Analysis, University of Plovdiv Paisii Hilendarski, 4000 Plovdiv, Bulgaria; kulina@uni-plovdiv.bg (H.K.); aivanov@uni-plovdiv.bg (A.I.)
* Correspondence: snow@uni-plovdiv.bg

Abstract: The aim of this study is to evaluate students' achievements in mathematics using three machine learning regression methods: classification and regression trees (CART), CART ensembles and bagging (CART-EB) and multivariate adaptive regression splines (MARS). A novel ensemble methodology is proposed based on the combination of CART and CART-EB models in a new ensemble to regress the actual data using MARS. Results of a final exam test, control and home assignments, and other learning activities to assess students' knowledge and competencies in applied mathematics are examined. The exam test combines problems on elements of mathematical analysis, statistics and a small practical project. The project is the new competence-oriented element, which requires students to formulate problems themselves, to choose different solutions and to use or not use specialized software. Initially, empirical data are statistically modeled using six CART and six CART-EB competing models. The models achieve a goodness-of-fit up to 96% to actual data. The impact of the examined factors on the students' success at the final exam is determined. Using the best of these models and proposed novel ensemble procedure, final MARS models are built that outperform the other models for predicting the achievements of students in applied mathematics.

Keywords: mathematical competency; assessment; machine learning; classification and regression tree; CART ensembles and bagging; ensemble model; multivariate adaptive regression splines; cross-validation

Citation: Gocheva-Ilieva, S.; Kulina, H.; Ivanov, A. Assessment of Students' Achievements and Competencies in Mathematics Using CART and CART Ensembles and Bagging with Combined Model Improvement by MARS. *Mathematics* **2021**, *9*, 62. https://doi.org/10.3390/math9010062

Received: 30 November 2020
Accepted: 26 December 2020
Published: 30 December 2020

Publisher's Note: MDPI stays neutral with regard to jurisdictional clai-ms in published maps and institutio-nal affiliations.

Copyright: © 2020 by the authors. Licensee MDPI, Basel, Switzerland. This article is an open access article distributed under the terms and conditions of the Creative Commons Attribution (CC BY) license (https://creativecommons.org/licenses/by/4.0/).

1. Introduction

The quality of mathematics training in higher education is essential for competitive future professional achievements of students in engineering, software, economics and other specialties. Alongside traditional teaching and learning methods in mathematics, increasingly various information technologies, computer- and mobile-based methods are applied using specialized software, as well as methodologies that involve project and team work, group discussions, role playing, blended learning etc. [1,2]. In particular, in the last two decades, an overall vision for teaching mathematical subjects in connection with their possible practical applications has been actively developed regarding the concept of competence. The concept is defined in [3] as follows: "Mathematical competency is understood as the ability to understand, judge, do, and use mathematics in a variety of intra- and extra-mathematical contexts and situations in which mathematics plays or could play a role."

In the context of higher education in engineering [3,4], the following eight key mathematical competencies are formulated: C1—thinking mathematically; C2—reasoning mathematically; C3—posing and solving mathematical problems; C4—modeling mathematically; C5—representing mathematical entities; C6—handling mathematical symbols and formalism; C7—communicating in, with, and about mathematics; C8—making use of aids and tools. Based on these competencies, specific teaching and learning methods for mathematics and assessment of student knowledge in engineering higher education and assessment standards were discussed in [5,6], which report the results of statistical modeling of data from a summative and competency-based assessment test using two machine learning and data mining techniques—cluster analysis and classification and regression trees (CART). With the aid of these methods, models are obtained for classification and determination of dependencies, for predicting student achievements based on the grades from a linear algebra and analytical geometry test and a short 10 min general mathematical competence test. A recent paper [7] analyzes the results from a written test of knowledge and the accompanying self-assessment survey of the examined students for individual problems using the CART method.

To improve and measure the level of student knowledge and competencies, in literature both traditional and modern cognitive and statistical approaches are applied. Standard multivariate statistical methods for the assessment of student knowledge are used, for example, in [8] to measure mathematical competencies of students upon admission to university with the help of Rash analysis and other analyses to provide insights into the measures' reliability and validity. A methodology for improving communication competencies and skills by learning mathematics in engineering degree specialties is presented with examples in [9]. Numerous publications use educational data mining (EDM) to establish classifications and dependencies in heterogeneous types of information related to education at all levels. EDM encompasses several research fields, such as data mining, machine learning (ML) and statistics. A recent review article [10] provides systematic information and analysis of a large number of studies, which use soft computing methods in EDM and ML for 2010–2018. The authors emphasize that decision tree, random forest, artificial neural network (ANN), fuzzy logic, support vector machine (SVM) and genetic/evolutionary algorithms are a few examples of soft computing approaches that, given enough data, can successfully deal with uncertainty, qualitatively stated problems and incomplete, imprecise or even contradictory data sets. These types of methods have a wide scope of application for research into various problems. Classifying and predicting students' academic success are carried out in [11] using several decision tree algorithms. A model is obtained which successfully predicts 79% of the grades of the students involved. To predict the student's performance in the Introduction to Informatics module, the authors in [12] applied six ML techniques, namely naïve Bayes, decision trees C4.5, NN, instance-based learning, logistic regression and SVM. It was found that the naïve Bayes algorithm is the most appropriate technique. In [13], decision trees, artificial neural networks and naïve Bayes models are built to predict students' academic performance based on their academic record, personal data and social information. Decision tree classification and regression models are built and studied for evaluation of mathematical competencies and student success in [7], achieving model performance of over 90% for both types of models. The highly effective data mining and ML technique, random forest (RF), is used in [14] for predicting students' dropout from university. In [15], a comparative study of seven predictive models for high school student performance in mathematics is performed using ML, deep learning and other techniques. The RF models show the best qualities with over 90% predictive performance. The same authors apply four ML techniques in [16] and also build hybrid models utilizing principal component analysis. The best predictive results of up to 98% with minimal relative error are obtained by RF models. SVM, ANN, fuzzy functions and other types of ML models are obtained and analyzed in several papers [17–21]. Other examples of ML methods with applications in education can be found in review papers [22–24]. Recent advances related to all kinds of ensemble learning algorithms, frameworks and methodologies, and their applications, can be found in [25].

The aim of this study is to demonstrate a combined traditional and competence-oriented approach to conducting an examination test in mathematics, as well as to determine factors affecting students' mathematics achievements and competencies using powerful predictive ML techniques. A case study is performed with results from the final exam in the course of Applied Mathematics with the first year students in specialty Business Information Technologies at University of Plovdiv Paisii Hilendarski, Bulgaria, which also includes as its principal component a small practical project. The main predictors used in the analyses are students' grades from ongoing testing during the trimester (control works and home assignments), attendance at lectures and laboratory practice, as well as the scores on individual problems in the exam and a small practical project. The modeling of the empirical data is performed with the methods CART and CART ensembles and bagging (CART-EB). To improve the result of the prediction of the exam test points, the best of these models are assembled with MARS.

This is the first time that the CART-EB method is applied for statistical modeling of data in the field of education. Another contribution is the use of the MARS method to generate new ensemble models from other ensemble models.

2. Materials and Methods

2.1. Methodology

The main part of each training process in education is the assessment of knowledge and skills, acquired by the students at a given stage. Depending on the curriculum for a given mathematical subject, in order to pass the exam, the student attends a certain number of lectures and laboratory practices, takes intermediate tests (control tests), solves assignments at home, works on individual or group projects, prepares presentations etc. Usually this type of control is evaluated with a certain score. This combination of activities is denoted as preparatory. At the end of their education, the students take a final exam, which can be written, oral or a combination of the two, or another type of assessment. It is assumed that the grade from the exam is influenced by the combination of preparatory activities during the course of the education. In order to apply a competency-oriented approach to the assessment of acquired knowledge and skills, a small practical-oriented project is used as a component of the final exam. All preparatory activities and the components of the exam test can be assigned a certain type of measurement and the respective dataset can be derived, where the grades are presented as variables. Exam grades can be considered to be a dependent or target variable, and the rest are predictors. Potential predictors are, for example, homework grades, course project grades, reports, gender of the student, the high school he or she graduated from.

Our experimental empirical study sets out to perform the following tasks:

- Construction of the integrated competency-based test for the final exam in mathematics;
- Construction, analysis and improvement of predictive models for evaluation of students' achievements using ML techniques;
- Application of the models for determining the importance of the preparatory activities and individual components of the exam to its assessment and, in particular, the importance of the project.

In essence, these tasks point to finding hidden similarities and patterns in the data using ML regression-type modeling techniques.

2.2. Machine Learning Methods Used for Statistical Analyses

The term ML (also referred to as learning analytics) denotes a class of methods and algorithms of artificial intelligence. Usually ML is used for classification and regression problems, and self-learning is achieved through various algorithms for cross-validation, improvement of model accuracy and fitting quality. This is achieved by combining features of computational statistics tools, numerical methods, optimization methods, probability theory, graph theory etc. ML methods are nonparametric and allow the detection of nonlinearities and relationships in the data without the need to model them explicitly; that is, they are data driven. Their core advantage is the generation of numerous distribution free and robust models, among which the most adequate and optimal model in a given sense can be selected. The following ML methods are widely used to model educational data: logistic regression, cluster analysis, decision trees (CART), support vector machines (SVM), multivariate adaptive regression splines (MARS), random forest (RF), neural networks (NN), fuzzy logic and others [10,24].

2.2.1. Classification and Regression Tree (CART)

The CART method [26] is a typical representative of decision tree algorithms and can be used for classification or regression. The main concept of the method is to classify the data from the training dataset through a recursive procedure into a binary tree structure with nodes. At each stage, the cases in the current node, called parent node, can be split into two child nodes according to the threshold value of some predictor variable. The predicted value in a terminal node is simply the average of the response values located in that node [24]. The threshold value is determined by a greedy algorithm, which checks all variables and their values so that the model minimizes the current selected type of summary error of the predicted values or other criteria in the terminal nodes of the tree. The splitting criterion for regression trees can be least squares or least absolute deviation. Once the tree is constructed, branches that do not contribute to the improvement of the model are removed and a final pruned tree is obtained. The researcher presets the settings and hyperparameters to select an optimal model, the type of cross-validation or other ML procedure, and adjusts the algorithm. For more details, see [27,28].

2.2.2. CART Ensembles and Bagging (CART-EB)

There are cases in which regression CART models may show instability in prediction under the influence of outliers, unsignificant predictors, predictors with small variation and others. There may also arise overfitting of the model [29]. Then, it is appropriate to use an ensemble of trees in combination with a bagging (also known as bootstrap aggregation) algorithm. There are many ML methods involving these techniques. In the current study, the algorithm of the CART-EB method was applied using the analysis engine CART ensembles and bagger of the Salford Predictive Modeler software suite [30]. Some other implementations in literature can be found in [31,32]. The initial CART tree of the ensemble is constructed with the entire data sample and all predictors. Then, it is pruned using 10-fold cross-validation. Bagged trees are built independently one from the other on bootstrap samples with or without repeated cases. They use a random subset of predictor variables at each decision split as in the RF algorithm. Ensemble trees can be built as exploratory (unpruned) maximal trees or they can be pruned by cross-validation. The case-predicted value is the average of the predictions of all the trees in the ensemble.

2.2.3. Multivariate Adaptive Regression Splines (MARS)

MARS is a nonparametric data mining and machine learning method, developed in [33]. If the dependent variable (here *Exam*) is $y = y(X)$ and $X = (X_1, X_2, \ldots, X_p)$ are p predictors with dimension n, the regression MARS model $\hat{y} = \hat{y}_{[M]}$ has the following form:

$$\hat{y}_{[M]} = b_0 + \sum_{j=1}^{M} b_j BF_j(X) \tag{1}$$

where $b_0, b_j, j = 1, 2, \ldots, M$ are the coefficients in the model, $BF_j(X)$ are its basis functions (BF), M is the number of BFs. The one-dimensional BF is written in the form

$$BF_j(X) = \max_{X_k}(0, X_k - c_{k,j}) \text{ or } BF_j(X) == \max_{X_k}(c_{k,j} - X_k, 0), \tag{2}$$

where the nodes $c_{k,j} \in X_k$ are determined by the MARS algorithm. For the nonlinear interactions, BFs are built as products of other BFs.

The control parameters chosen by the researcher are the maximum number of basis functions and the maximum number of their multipliers (i.e., degree of interactions) in BFs. The algorithm involves two steps. The first step starts by setting b_0 (for example, $b_0 = \min_{1 \leq i \leq n} y_i$) and then the model is complemented consistently by BFs of type (2). For each model with a given number of BFs, the MARS algorithm defines variables and nodes so as to minimize a predefined loss function, such as the root mean square error. In the second step, BFs that do not contribute significantly to the accuracy of the model are removed. For more details, see [33].

2.2.4. Model Evaluation Metrics

In this study, the best ML models were selected using the highest coefficient of determination R^2 and the minimum values of the root mean square error (RMSE) given by the expressions

$$R^2 = \frac{\sum_{i=1}^{n}(\hat{y}_i - \bar{y})^2}{\sum_{i=1}^{n}(y_i - \bar{y})^2}, \quad RMSE = \sqrt{\frac{1}{n}\sum_{i=1}^{n}(y_i - \hat{y}_i)^2} \quad (3)$$

where $\hat{y}_i$ and y_i stand for model predicted and *Exam* values, respectively.

The performance of the models was also evaluated using the Theil's forecast accuracy coefficient U_{II} [34]:

$$U_{II} = \sqrt{\frac{\sum_{i=1}^{n}(y_i - \hat{y}_i)^2}{\sum_{i=1}^{n} y_i^2}}. \quad (4)$$

The lower the value of the coefficient, the better the accuracy of the model. The coefficient U_{II} is dimensionless and is used to compare models obtained by different methods, as well as to identify large values. The model is considered to be of good quality when (4) is less than 1.

When choosing from nested models, the parsimony principle was applied [35].

3. Results

3.1. Test Design

An experiment was conducted with the Applied Mathematics course. The final exam test combines three main components with problems in mathematical analysis, probability theory and applied statistics. It includes:

- Problems in math analysis (5 problems),   15 points, 50%;
- Problems in probability (2 problems),   5 points, 17%;
- A small practical project in applied statistics,   10 points, 33%.

The percentage indicates the relative weight within the total number of 30 points for the entire exam. Unsolved problems are evaluated with 0 points. A sample version of the exam test with 7 type variations is given in Figure 1. Each student works on an individual test. It needs to be noted that the problems in the first two components are of traditional type; these problems have been used in exams in this course of studies over the last 7–8 years. The added project includes some general instructions without explicitly stating how the problem is to be solved.

The exam was taken by 68 first-year students in the specialty of Business Information Technologies at the Faculty of Mathematics and Informatics, University of Plovdiv Paisii Hilendarski, Plovdiv, Bulgaria. According to the first trimester curriculum, these students have taken a linear algebra and analytic geometry course, and during the second trimester, the course in Information Technology for Mathematics, where students are trained to use Wolfram Mathematica to solve mathematical problems using a computer. The current course in Applied Mathematics is in the third trimester.

The results of the preliminary activities and the final exam in number of points are described with the variables: *Exam* (total exam points, up to 30), *Math_An* (mathematical analysis, up to 15), *Stat* (statistics, up to 5), *Project* (up to 10), *A1_12* (home assignment 1, up to 12), *A2_20* (assignment 2, up to 20), *CW1_30* (homework 1, up to 30), *CW2_30* (homework

2, up to 30), *Attn_Lect* (attendance to lectures, up to 10) and *Attn_Labs* (attendance to labs, up to 10).

Exam in Applied Mathematics for the specialty Business Information Technologies, 1ˢᵗ year

Substitute: *p* = penultimate digit of faculty number +1; *q* = last digit of faculty number +1

Problem 1 (2 + 2 + 3 + 3 + 5 points). The function $f(x) = \dfrac{px + q}{x^2 - 9}$ is given.

A) Determine the definition domain of $f(x)$.
B) Find the first derivative of $f(x)$.
C) According to the sign of the first derivative to find the intervals of increase and decrease of the function. You can draw the necessary graphs.
D) Determine the limits $\lim\limits_{x \to -3} f(x)$ and $\lim\limits_{x \to 3} f(x)$.
E) Calculate the area locked between the function and the abscissa axis for $x \in [0,1]$.

Problem 2 (3 + 2 points). Gymnast trains a combination with bats. In the basket are collected 6 pairs of new and 5 pairs of old bats. If she plays with new bats, the probability of dropping a bat is $p/20$, and if she plays with old bats, the probability is $(p+q)/20$.

A) Find the probability of dropping a bat if the gymnast chooses any pair of bats from the basket.
B) Find the probability of not dropping a bat if she has chosen a new pair.

Problem 3 (Small Project, 10 points). Load the Real_estate3.sav file with SPSS. Delete the first 5 $(p + q)$ lines. The file contains data on houses for sale in a large city. The variables are:
No - number of the case, X1 transaction date, X2 house age, X3 distance to the nearest metro station, X4 number of convenience stores, X5 latitude, X6 longitude, Y house price of unit area.

A) Calculate the descriptive statistics of the variables. Check the type of distribution.
B) Use appropriate statistical analyses to investigate whether there is a significant dependence in the dataset.
C) Remove the insignificant variables and repeat the analysis. Build the model to identify the factors that affect the price and interpret the result.
D) What percentage of the data explains the model?

The decision should be explained on a separate sheet.

Total: maximum 30 points.
Send the solutions with the Wolfram Mathematica and SPSS working files, to the teacher's e-mail.

Figure 1. Example of the exam test in Applied Mathematics.

3.2. Measurement of Competencies by the Exam Test

Following the recommendations of [4], the experience in [5,36] and with the aid of a three-dimensional scale, we defined the correspondence between the eight competencies and the elements of the exam test, as shown in Table 1. Here T1–T5 mean subproblems A–E in problem 1; S1 and S2 in problem 2; P1–P4, the instructions to the project. It is shown that all competencies are included with the exception of C7 because the exam is individual and does not allow for any communication with other students and/or external sources. In addition, Figure 1 shows that the probability theory problem *Stat* requires a solution with pen and paper and does not duplicate the project. As a whole, the project is independent in terms of curriculum covered and supplements the competencies, which are not included in the first two test components. The level of solution of the project indicates the degree to which the students have acquired the necessary knowledge and skills in statistics in order to solve on their own a complete mathematical problem—from the data, through the analyses to the interpretation of the results obtained.

It should be noted that the students have solved the project in different ways, with different methods. Some managed to make only descriptive statistics, with different statistics selected. Other students continued with cluster analysis, factor analysis or principal component analysis. More often, regression analysis was performed in one-dimensional or multidimensional case.

Table 1. Assessments of the level of mathematical competencies in problems from the *Exam* test [1].

	Competency	Exam Elements									
		T1	T2	T3	T4	T5	S1	S2	P1	P2	P3	P4
C1	Thinking mathematically	−	−	−	−	0	+	+	0	0	0	0
C2	Reasoning mathematically	−	−	0	−	0	0	0	−	0	0	−
C3	Problem solving	0	0	0	0	+	+	+	0	0	0	−
C4	Modeling mathematically	−	−	−	−	−	−	−	+	+	+	0
C5	Representation	0	0	+	+	0	+	−	−	−	−	−
C6	Symbols and formalism	0	0	0	0	0	−	−	0	0	0	−
C7	Communication	−	−	−	−	−	−	−	−	−	−	−
C8	Aids and tools	−	0	−	+	+	−	−	+	+	+	−

[1] The meaning of the signs: +, "very important"; 0, "medium important"; −, "less important".

3.3. Initial Processing and Analysis of the Data

Descriptive statistics of the variables used in the study are given in Table 2 and the distributions in the form of box plots with unstandardized data are shown in Figure 2a,b. Table 2 and Figure 2b show that the mean values of the results for *Stat* and *Project* are quite low, and their median is 0. The reason is because only 31 students, or 45%, worked on the project. In addition, Table 2 and Figure 2a,b lead us to the conclusion that most of the variables are not normally distributed (A2_20, CW_30, Stat, Project etc.). This is evidenced by the relatively high values of the ratios of skewness/std. error of skewness, kurtosis/std. error of kurtosis, as well as from the box plots. For example, for the target variable *Exam* we have the ratios $\frac{Skewness}{Std.\ Err.\ Skewneess} = \frac{1.14}{0.29} = 3.931 > 1.96$ and $\frac{Kurtosis}{Std.\ Err.\ Kurtosis} = \frac{1.85}{0.57} = 2.643 > 1.96$, which is an indication for non-normal distribution of the variable. In addition, a one-sample Kolmogorov–Smirnov test with Lilliefors significance correction was applied, which reaffirms that *Exam* did not follow a normal distribution as the calculated p-value is 0.000. Furthermore, the relationships between the variables are hidden and possibly highly nonlinear.

Table 2. Descriptive statistics of the initial predictors and the target variable.

Statistics	Attn_ Lect	Attn_ Labs	A1_ 12	A2_ 20	CW1_ 30	CW2_ 30	Math_ An	Stat	Project	Exam
Mean	6.87	5.63	6.65	12.46	10.07	7.88	8.67	1.76	2.48	12.91
Median	8.00	5.00	7.00	15.75	9.25	6.00	8.75	0.00	0.00	12.00
Std. Deviation	3.44	3.11	3.71	6.94	8.35	7.24	3.64	2.234	3.18	4.88
Variance	11.82	9.67	13.75	48.13	69.73	52.46	13.25	4.99	10.12	23.78
Skewness	−0.81	0.19	−0.32	−0.90	0.51	0.44	0.12	0.81	0.86	1.14
Std. Error of Skewness	0.29	0.29	0.29	0.29	0.29	0.29	0.29	0.29	0.29	0.29
Kurtosis	−0.72	−1.38	−0.85	−0.79	−0.68	−1.12	−1.018	−0.91	−0.79	1.85
Std. Error of Kurtosis	0.57	0.57	0.57	0.57	0.57	0.57	0.57	0.57	0.57	0.57
Range	10	10	12	19.5	30.0	22	2.0	0.00	10.0	25.0
Minimum	0	0	0	0.0	0.0	0	15.0	7.00	0.0	4.0
Maximum	10	10	12	19.5	30	22	8.7	1.76	10.0	29.0

3.4. Results from the CART Models

Multiple regression trees were built using the CART method. The dependent variable is *Exam* and the factors on which its values depend are the remaining nine variables, i.e., *Math_An, Stat, Project, A1_12, A2_20, CW1_30, CW2_30, Attn_Lect,* and *Attn_Labs*. The objective was to define which independent factors have the strongest influence on the *Exam* and to what extent. Before applying the algorithm, hyperparameters m1 (minimum cases in parent node) and m2 (minimum cases in child node) were set. Regression tree procedure on the learn (training) set is carried out using 10-fold cross-validation, which is recommended for small samples [27,28]. The least squares method was selected as a splitting criterion.

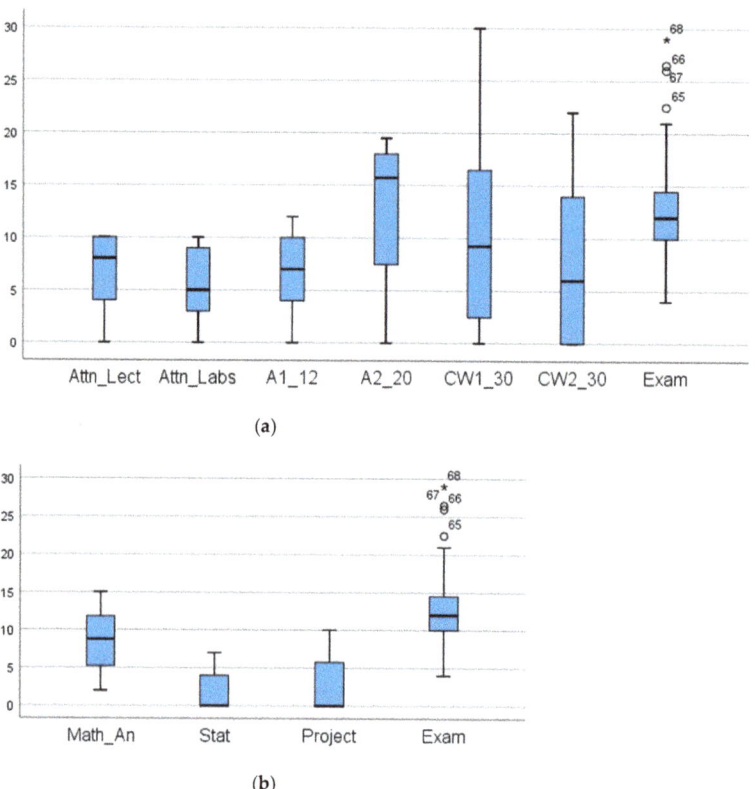

Figure 2. Box plots of the initial predictors and target variable *Exam*, used in the statistical analyses (nonstandardized): (**a**) preliminary activities and *Exam*, (**b**) *Exam* elements.

For m1 = 5, m2 = 2 in Figure 3a, a diagram is shown of the relative error of the generated CART models depending on the number of their terminal nodes. For the case of m1 = 6, m2 = 3, the scheme with the relative errors of the constructed models is presented in Figure 3b. Relative errors are calculated as the ratio of the least square error of the current model divided by the root node error. Models with relative errors distinguished by one standard error (1 SE) are colored in green. This means that all models in green from Figure 3a,b can be considered as a set of competing models. From Figure 3a, it is evident that the model with a minimum relative error of 0.321 has 13 terminal nodes. We denote it by M_1. In addition, two models were analyzed—the M_2 model with a minimum number of 9 nodes and the maximum M_3 model with 22 nodes. Besides these, in the same way we denote the model with a minimum relative error of 0.310 and 11 terminal nodes with M_4, the model with 9 terminal nodes with M_5, and the model with 19 terminal nodes with M_6, respectively.

Table 3 contains summary statistics for the competing six CART models $M_1, M_2, \ldots, M_6$ that are selected. We compare the two optimal models M_1 and M_4. Although model M_4 has larger constraints of m1 = 6 and m2 = 3, it shows the highest value of R^2 test = 0.698, and the minimum value of RMSE Test = 2.694. At the same time, this model is inferior to the prediction statistics, especially with the relatively large RMSE = 1.517, which is 16% higher than that of M_1. From the first group of "finer" models, M_1 is comparable to

M_4 with R^2 Test = 0.621 compared to 0.698 for M_4 (1%), RMSE Test = 2.743, with a small difference of 0.049, or less than 2%. The goodness-of-fit R^2 Learn = 0.928 of the M_1 model is 3% higher than the M_4 statistic (0.902). Next, we make a comparison between M_1 and M_6. The statistics of these two models are almost identical, but the M_6 model is more complex as it contains 19 terminal nodes compared to the 13 of M_1. From the set of competing models considered, we should reject M_2 and M_5. This is due to the most unsatisfactory statistics—the smallest R^2 and the largest RMSE Learn of the prediction. Model M_3 has a less favorable relative error compared to M_1 (with 4%), outperforming M_1 narrowly for the main indicators (4) by 1 to 9%. Since the M_3 model is the most complex, having 22 terminal nodes, compared to the 13 of M_1, we have to apply the parsimony principle [34,35] (see also Figure 3a). We will further consider CART models M_1 and M_4. Note that all Theil's coefficients are sufficiently small.

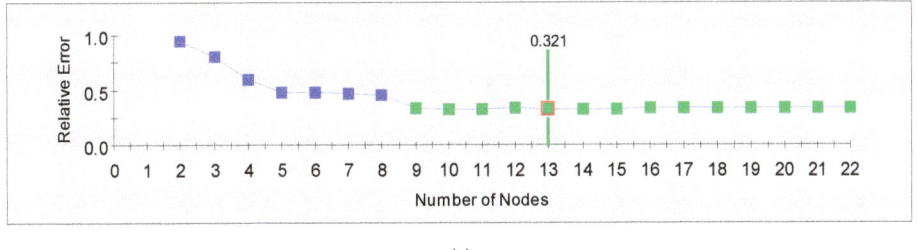

(a)

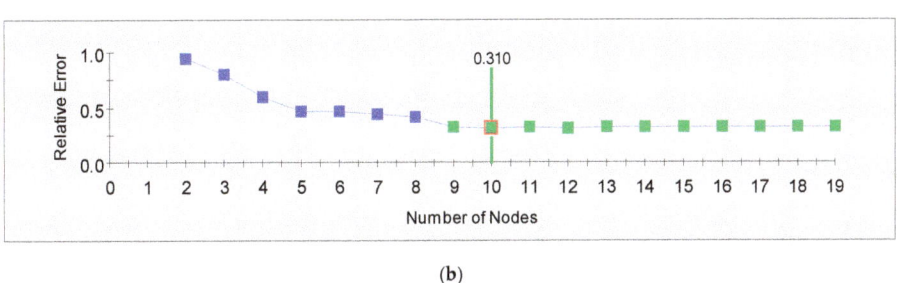

(b)

Figure 3. Relative error of the constructed classification and regression tree (CART) models depending on the number of terminal nodes: (a) m1 = 5, m2 = 2; (b) m1 = 6, m2 = 3.

Table 3. Summary statistics of the selected regression CART models for assessment of students' achievements.

Statistic	Model					
	M_1	M_2	M_3	M_4	M_5	M_6
Terminal nodes	13	9	22	10	9	19
m1-m2	5-2	5-2	5-2	6-3	6-3	6-3
Relative error	0.321	0.333	0.335	0.310	0.319	0.325
R^2, Test	0.621	0.683	0.678	0.698	0.690	0.685
R^2, Learn	0.928	0.892	0.940	0.902	0.899	0.929
RMSE, Test	2.743	-	-	2.694	-	-
RMSE, Learn	1.298	1.588	1.188	1.517	1.616	1.291
Theil's U_{II}	0.0089	0.0133	0.0074	0.0121	0.0137	0.0088

Table 4 presents the values of the relative importance of the nine factors studied on the *Exam* points according to their participation in the exploratory Learn CART trees. Here, too, stability is clearly visible, with small differences. The largest factor of importance

(100 relative points) is the *Project* factor—the points from the exam obtained for solving the small practical project. The next main factors, in descending order of their relative importance, are the scores on *Math_An* (93–95 relative units) and *A2_30* (50–54 units). The points from solving the problems in the *Stat* statistics have a small impact, within 20–22 relative units. Table 4 also shows that the influence of the predictors obtained in the chosen optimal model M_1 is almost identical to that of model M_3, as well as the other maximal model M_6.

Table 4. Relative importance of the initial predictors used in the selected CART models.

Predictors	CART Models					
	M_1	M_2	M_3	M_4	M_5	M_6
Project	100	100	100	100	100	100
Math_An	95.25	93.76	95.27	93.25	93.08	94.15
A2_20	54.34	50.52	54.59	52.80	50.52	54.17
CW1_30	37.79	35.99	38.63	36.36	35.99	37.97
A1_12	36.02	35.10	36.29	35.29	35.29	36.68
Attn_Labs	34.15	34.27	34.45	34.27	34.27	33.90
CW2_30	22.66	19.72	23.31	22.12	19.72	22.84
Stat	21.74	20.28	21.57	22.04	20.28	21.58
Attn_Lect	6.23	5.03	6.90	6.26	5.03	6.39

The calculation of the coefficients of importance in Table 3 is performed using sequential aggregation. At level 0, the mean value of target (*Exam*) as predicted by the model is calculated for the entire sample and the RMSE is calculated. At the first split (as shown in Figure 4), the CART algorithm selected the *Math_An* as a splitter predictor and its threshold value is 11.25. After the split, the predictions (mean values) are calculated in both child nodes along with their RMSEs. The relative improvement of the current accuracy achieved is calculated against the root, and the value obtained is added to the coefficient of importance of the predictor. The process is repeated until the tree is built.

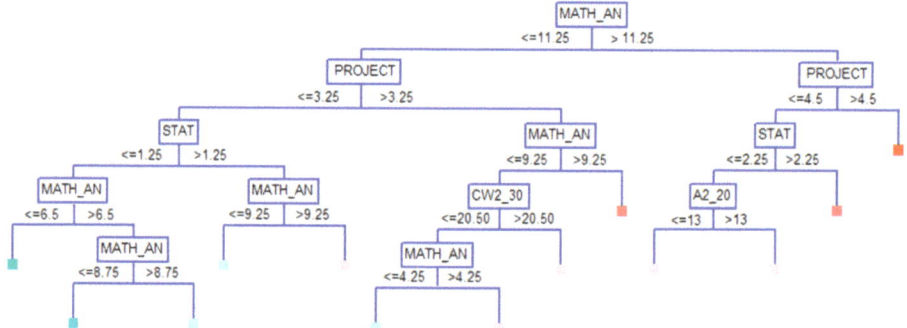

Figure 4. Diagram of the regression tree of CART model M_1 with split variables denoted and respective threshold values for each tree node.

Figure 4 presents the regression tree of model M_1 with the splitting variables and their threshold values at each split. The initial splitting variable at the tree root is *Math_An*. The rule for splitting each root case into two is "cases with a value of *Math_An* <= 11.25 go into the left child node and the rest into the right one". Here the threshold value is *Math_An* = 11.25. At the next (first) level, the splitting variable for both nodes is *Project* with the same splitting rule for the cases: "*Project* <= 4.5" etc. The process which generates the CART tree M_1, shown in Figure 4, is stopped to a depth 5, having 13 terminal nodes, marked with a colored square. The value predicted by the model for each case of the given terminal node is the arithmetic mean of *Exam* points from the cases classified in that node.

The tree of the M_4 model shown in Figure 5 has an identical structure as the tree of the M_1 model from Figure 4. An almost complete match is observed; therefore, stability of the CART models is obtained.

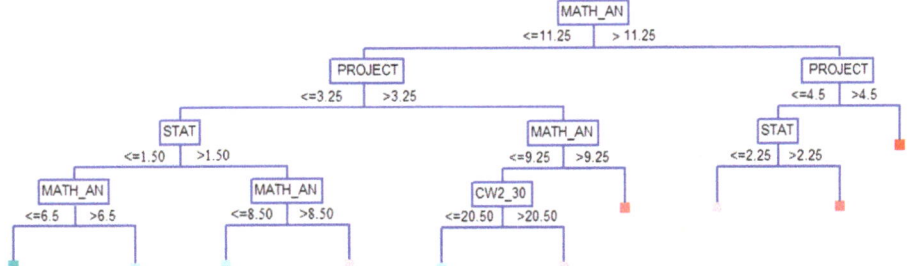

Figure 5. Diagram of the regression tree of CART model M_4 with split variables denoted and respective threshold values for each tree node.

The predicted scores obtained from the six CART models in Table 4 were statistically examined with a Wilcoxon signed rank test for paired samples. It was found that all Wilcoxon signed rank tests were statistically unsignificant and the differences of each two models had symmetrical distributions. This is an indicator that the models do not differ significantly from each other.

Figure 6 shows the actual versus predicted values of the *Exam* obtained by model M_1.

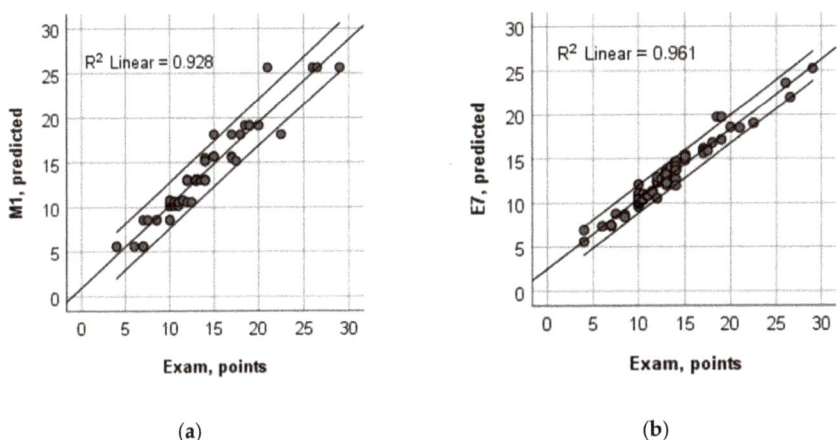

Figure 6. Scatter plots of predicted versus actual values of the target variable *Exam* with 5% confidence interval for (a) CART model M_1 and (b) CART-EB model E_7.

Figure 7 presents line plots of *Exam* and model predictions obtained by models M_1 and E_7. For both models there are larger differences with *Exam* at the highest values.

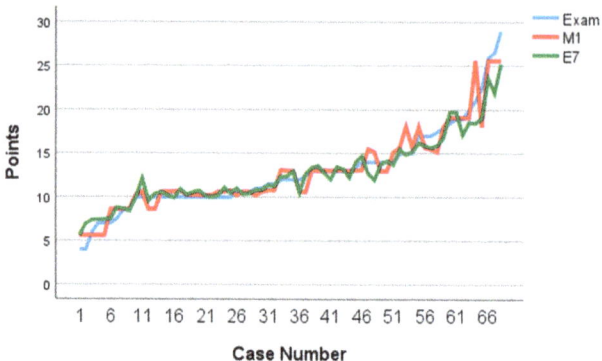

Figure 7. Line plots of the target variable *Exam* and its predicted values with models M_1 and E_7.

3.5. Results from the CART Ensembles and Bagged Models

The same values of the hyperparameters as in the CART models were used in the construction of the CART-EB models. For the minimum number of cases in a parent node (m1) and the minimum number of cases in a child node (m2), two options were set: 5-2 and 6-3, respectively. All trees in the ensemble were trained with 10-fold cross-validation. Bagged trees were built with repeated cases. Due to the small sample size ($n = 68$), the ensembles were built with 10, 15, 20 and 25 trees. The family of these models are denoted by $E_1, E_2, \ldots, E_8$. Of these, the first group E_1, E_2 and E_3 are built using values of the hyperparameters m1-m2 equal to, respectively, 6-3. For the second group with the remaining models, these parameters are 5-2. The first models of each group, namely E_1 and E_4, are initial CART trees. Table 5 presents the statistical indicators obtained for these models. The models E_2 and E_3 show relatively low R^2 Test and the largest errors RMSE Test and Learn. Also the corresponding Theil's coefficients are larger than those of the models with m1 = 5 and m2 = 2. Model E_6 has the best test statistics with R^2 Test (0.922) and RMSE Test (1.838). The next model E_7 has the best indicators for the training sample—R^2 Learn (0.961) and RMSE Learn (1.278). As the number of terminal nodes increases, the statistics become less favorable, as seen from those of model E_8. Therefore, in further analysis we consider the models E_6 and E_7. The predictive properties of model E_7 are illustrated in Figures 6b and 7.

Table 5. Summary statistics of regression CART ensembles and bagged models for assessment of students' achievements.

Statistic	CART Ensembles and Bagged Model							
	E_1 Initial Tree	E_2	E_3	E_4 Initial Tree	E_5	E_6	E_7	E_8
Number of trees	-	10	15	-	10	15	20	25
m1-m2	6-3	6-3	6-3	5-2	5-2	5-2	5-2	5-2
R^2, Test	0.845	0.886	0.897	0.845	0.916	0.922	0.883	0.807
R^2, Learn	-	0.923	0.942	-	0.936	0.953	0.961	0.945
RMSE, Test	2.400	2.222	2.055	2.368	2.092	1.838	1.908	2.302
RMSE, Learn	-	1.724	1.583	-	1.610	1.370	1.278	1.368
Theil's U_{II}	-	0.0121	0.0132	-	0.0177	0.0104	0.0086	0.0098

The results of the Wilcoxon signed rank tests show that the built CART-EB models do not differ significantly from each other.

3.6. Combination of CART and CART Ensembles and Bagged Models Using MARS

To improve the quality of prediction, MARS regression models of the dependent variable *Exam* were built using the selected four best models M_1, M_4, E_6 and E_7 as predictors. Due to the almost linear behavior of the *Exam* curve, only a linear MARS method was applied. The MARS models generated are denoted by MM_1, MM_2 and MM_3. Furthermore, by finding the importance of individual regression trees models, it can be determined which of them has the best predictive properties. The models were trained with 10-fold cross-validation. The summary statistics of the models obtained are presented in Table 6.

Table 6. Summary statistics of regression of MARS models built using CART and CART-EB models as predictors [1].

Statistic	MARS Model		
	MM_1	MM_2	MM_3
Predictors	M_1, E_7	M_1, E_6, E_7	M_1, M_4, E_6, E_7
Number of BFs	4	5	6
Variable importance [2]	23, 100	42, 29, 100	97, 0, 55, 100
R^2 Test	0.960	0.956	0.954
R^2 Learn	0.972	0.974	0.978
GCV R^2	0.958	0.960	0.960
RMSE Test	0.966	1.021	1.035
RMSE Learn	0.804	0.749	0.726
Theil's U_{II}	0.0034	0.0029	0.0028

[1] GCV R^2 stands for general cross-validation R^2 [30,33]. [2] Variable importance corresponds to the order of the predictors in column 4.

The statistics of the built models in Table 6 are almost the same. Using all four models (see MM_3), we found that the greatest influence was exerted by model E_7 (100 relative units), followed by M_1 (97 units), E_6 (55 units) and M_4 (0 units). By successively reducing the predictors, the other models are obtained. We choose the simplest MM_1 model for optimal. MM_1 outperforms the separately taken CART and CART-EB models in all evaluation metrics from (4). The MM_1 model has the form

$$\hat{Exam} = 3.64249 + 1.52178 * BF1 + 0.275286 * BF3 - 0.843117 * BF5 + 0.321376 * BF7,$$
$$BF1 = max(0, E_7 - 5.605), \quad BF3 = max(0, E_7 - 14.1583), \tag{5}$$
$$BF5 = max(0, E_7 - 9.5625), \quad BF7 = max(0, M_1 - 10.1875).$$

Figure 8 shows the scatter plot of the actual *Exam* values versus MM_1 model predictions.

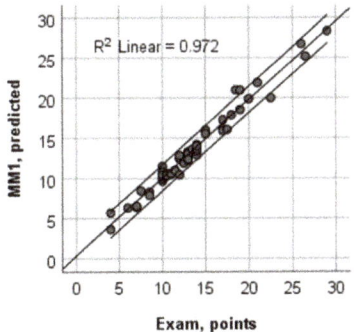

Figure 8. Scatter plot of the predicted values from optimal MARS model MM_1 versus actual values of the target variable *Exam* with a 5% confidence interval.

4. Discussion with Conclusions

This study presents, models and analyzes results from a competency-based exam in applied mathematics together with the results of preparatory academic activities. They are modeled using three ML methods—CART, CART ensembles and bagging, and MARS.

The CART method was first applied. Table 3 indicates that the six CART models selected have high goodness-of-fit indicators with coefficients of determination R^2 over 90% and RMSE around 1.5, or within 5%. As optimal models, we selected M_1 and M_4. M_1 shows $R^2 = 0.928$, RMSE = 1.298, as well as a small value of the Theil's forecast accuracy coefficient $U_{II} = 0.0089$.

The CART models allow for determining the influence of individual educational components on exam results for the specific subject of applied mathematics. The importance of the predictors in the M_1 model in relative units is *Project* (100), *Math_An* (95), *A2_20* (54), *CW1_30* (38), *A1_12* (36) etc., as presented in Table 4. Therefore, the solution of the project and the tasks of mathematical analysis determine to the greatest extent the achievements of the students. Other important factors are the grades from the second homework, the first control test etc. The influence of students' success with problem 2 in statistics (variable *Stat*) reaches only about 22% relative weight. This indicates an unsatisfactory level of theoretical knowledge in probabilities and statistics. Using these results with Table 1 of competencies, it is apparent that following a reduction of competencies for the two problems from *Stat*, the exam test can be assessed mainly by the acquisition of competencies with the most "+" and "0". These competencies are C3, C4, C6 and C8.

The data were also modeled using the CART ensembles and bagging method. Six CART-EB models were built. The analysis of these models showed that the best statistical evaluation indicators were for models E_6 and E_7, with 15 and 20 trees in the ensemble, respectively. As an optimal model, E_7 achieved $R^2 = 0.961$, RMSE = 1.278, as well as a Theil's accuracy coefficient $U_{II} = 0.0086$. Although the EB models do not derive the influence of individual predictors, they serve as confirmatory and complementary to CART models for predicting student achievement.

The idea arose to combine the four best models—two CART and two CART-EB models for regression of the dependent variable *Exam*. A linear MARS method was applied. Three models with predictors M_1, M_4, E_6 and E_7 were built. The models showed very close goodness-of-fit indices. The final model selected, MM_1, used M_1 and E_7 and achieved $R^2 = 0.972$, RMSE = 0.804 and Theil's accuracy coefficient $U_{II} = 0.0034$. This model showed a significant improvement in the prediction of the lowest and highest exam scores.

The results obtained are comparable to those obtained by us in [6], where regression-type CART models were constructed to predict the final exam results in linear algebra and analytical geometry for students in two other specialties at the same university, using a short mathematical competency test and mid-term test results. In [6], the models reached to fit the actual data with $R^2 = 93\%$. The results obtained here are also similar to those in [11,13].

It should be noted that for the first time the CART ensemble and bagging method for data from education was applied. In addition, for the first time, combining the predictions of the individual competing models using the MARS method is used. The combined MARS models obtained exceed the qualities of the predictors included in all statistical indicators.

Essentially, the approach for modeling the data we present consists of two consecutive steps: (1) building regression trees and regression trees ensemble models (Sections 3.4 and 3.5) using the initial predictors, and (2) building MARS models (Section 3.6), where the predictors are the resulting variables with values predicted in step (1). Based on this, we can formulate some advantages and potential capabilities of this approach, namely:

- As part of the family of regression trees, the CART and CART-EB methods we use can successfully deal with uncertainty, qualitatively stated problems and incomplete, imprecise or even contradictory data sets, as stated in [10]. These can process both nominal and numerical data, handle multidimensional and multivariety data, easily identify patterns and nonlinear complex relationships between the predictors, thus facilitating the interpretation of models.

- At step (1), the variable importance of initial predictors in models is assessed directly, which allows us to ignore/screen unsignificant predictors. This would be especially useful in the case of a large number of predictors for reducing the dimensionality of the problem.
- At step (2), numerical-type data are used, enabling the implementation of the MARS method, whereby it is combined with the predictions from (1). In particular, the results of our study showed that MARS models improve the predictions of the smallest and largest values of the target variable, including its outliers. In this manner, it is possible to eliminate or reduce the effect of this type of flaw, typical for all ensemble methods.
- The importance of the models from step (1), used as predictors, is determined with the help of MARS in step (2). In this manner, the best regression trees model is identified. Indirectly, if it determines the influence of the initial predictors, additional useful information may be obtained to interpret the overall statistical analysis.

In addition to this, the proposed research method has several limitations. The models can be built if at least 50 data records are available. Furthermore, the CART-EB algorithm used does not deduce the relative importance of the predictors in the model, which makes any direct interpretation of the results difficult. Another disadvantage typical for all ML methods is that results depend to a certain degree on accuracy criterion, variable and model selection.

The proposed methods and models in this study can be used to direct and improve exam tests for students in subsequent courses, making changes at the tutor's discretion. Changes can be made both in the educational content, tests and the academic programs, and the management of other basic factors that influence grades, as determined using the models. This approach promises to find hidden relationships between factors contributing to learning and teaching, and also benefits tutors/authorities by making predictions and helping them make better decisions. Future research can be planned in this regard. By applying the approach we propose, another crucial practical issue for further research is determining the factors and predicting which students may drop out.

We can conclude that the use of small practical projects as a competency-oriented approach and combined with the application of powerful ML methods for processing the data set related to the learning process are effective for assessment of students' knowledge and competencies in mathematics.

Author Contributions: Conceptualization and methodology, all authors; data preparation, H.K. and A.I.; modeling, S.G.-I.; validation, H.K. and A.I.; analysis of the results, all authors; writing—review and editing, S.G.-I. All authors have read and agreed to the published version of the manuscript.

Funding: This work was accomplished with the financial support of the MES by the Grant No. D01-271/16.12.2019 for NCDSC part of the Bulgarian National Roadmap on RIs, financed by the Bulgarian Ministry of Education and Science.

Conflicts of Interest: The authors declare no conflict of interest.

References

1. Abdulwahed, M.; Jaworski, B.; Crawford, A. Innovative approaches to teaching mathematics in higher education: A review and critique. *Nord. Stud. Math. Educ.* **2012**, *17*, 49–68. Available online: https://repository.lboro.ac.uk/articles/Innovative_approaches_to_teaching_mathematics_in_higher_education_a_review_and_critique/9370940/files/16981556.pdf (accessed on 17 November 2020).
2. Hassan, O.A.B. Learning theories and assessment methodologies—An engineering educational perspective. *Eur. J. Eng. Educ.* **2011**, *36*, 327–339. [CrossRef]
3. Niss, M. Mathematical Competencies and the Learning of Mathematics: The Danish KOM Project. In Proceedings of the 3rd Mediterranean Conference on Mathematical Education, Athens, Greece, 3–5 January 2003; The Hellenic Mathematical Society: Athens, Greece, 2003; pp. 115–124. Available online: http://www.math.chalmers.se/Math/Grundutb/CTH/mve375/1112/docs/KOMkompetenser.pdf (accessed on 17 November 2020).

4. Alpers, B.A.; Demlova, M.; Fant, C.-H.; Gustafsson, T.; Lawson, D.; Leslie Mustoe, L.; Olsen-Lehtonen, B.; Robinson, C.; Velichova, D. *A Framework for Mathematics Curricula in Engineering Education: A Report of the Mathematics Working Group*; European Society for Engineering Education (SEFI): Brussels, Belgium, 2013; Available online: http://sefi.htw-aalen.de/Curriculum/Competency%20based%20curriculum%20incl%20ads.pdf (accessed on 17 November 2020).
5. Queiruga-Dios, A.; Hernández Encinas, A.; Demlova, M.; Dias Rasteiro, D.; Rodríguez Sánchez, G.; Sánchez Santos, M.J. Rules_Math: Establishing Assessment Standards. In *Advances in Intelligent Systems and Computing*; Martínez Álvarez, F., Troncoso, L.A., Sáez Muñoz, J., Quintián, H., Corchado, E., Eds.; Springer: Cham, Switzerland, 2020; Volume 951, pp. 235–244. [CrossRef]
6. Gocheva-Ilieva, S.; Teofilova, M.; Iliev, A.; Kulina, H.; Voynikova, D.; Ivanov, A.; Atanasova, P. Data Mining for Statistical Evaluation of Summative and Competency-Based Assessments in Mathematics. In *Advances in Intelligent Systems and Computing*; Martínez Álvarez, F., Troncoso, L.A., Sáez Muñoz, J., Quintián, H., Corchado, E., Eds.; Springer: Cham, Switzerland, 2020; Volume 951, pp. 207–216. [CrossRef]
7. Ivanov, A. Decision trees for evaluation of mathematical competencies in the higher education: A case study. *Mathematics* **2020**, *8*, 748. [CrossRef]
8. Neumann, I.; Rosken-Winter, B.; Lehmann, M. Measuring mathematical competences of engineering students at the beginning of their studies. *Peabody J. Educ.* **2015**, *90*, 465–476. [CrossRef]
9. Georgieva, P.V.; Nikolova, E.P. Enhancing communication competences through mathematics in engineering curriculum. In Proceedings of the 42nd International Convention on Information and Communication Technology, Electronics and Microelectronics, MIPRO 2019, Opatija, Croatia, 20–24 May 2019; Volume 8757207, pp. 1451–1456. [CrossRef]
10. Charitopoulos, A.; Rangoussi, M.; Koulouriotis, D. On the use of soft computing methods in educational data mining and learning analytics research: A review of years 2010–2018. *Int. J. Artif. Intell. Educ.* **2020**, *30*. [CrossRef]
11. Mesarić, J.; Šebalj, D. Decision trees for predicting the academic success of students. *Croat. Oper. Res. Rev.* **2016**, *7*, 367–388. [CrossRef]
12. Kotsiantis, S.; Pierrakeas, C.; Pintelas, P. Predicting students' performance in distance learning using machine learning techniques. *Appl. Artif. Intell.* **2004**, *18*, 411–426. [CrossRef]
13. Mueen, A.; Zafar, B.; Manzoor, U. Modeling and predicting students' academic performance using data mining techniques. *Int. J. Mod. Educ. Comp. Sci.* **2016**, *8*, 36–42. [CrossRef]
14. Behr, M.; Giese, M.; Teguim, K.; Theune, K. Early prediction of university dropouts—A random forest approach. *J. Econ. Stat.* **2020**, *240*, 743–789. [CrossRef]
15. Sokkhey, P.; Okazaki, T. Comparative study of prediction models for high school student performance in mathematics. *IEIE Trans. Smart Process. Comput.* **2019**, *8*, 394–404. [CrossRef]
16. Sokkhey, P.; Okazaki, T. Hybrid machine learning algorithms for predicting academic performance. *Int. J. Adv. Comput. Sci. Appl.* **2020**, *11*, 32–41. [CrossRef]
17. Qiang, T. Data mining algorithm and the effectiveness of mathematics classroom teaching based on support vector machine. *Int. J. Database Theory Appl.* **2016**, *9*, 163–174. [CrossRef]
18. Siri, A. Predicting students' dropout at university using artificial neural networks. *Ital. J. Soc. Educ.* **2015**, *7*, 225–247. [CrossRef]
19. Mat, U.B.; Buniyamin, N. Using neuro-fuzzy technique to classify and predict electrical engineering students' achievement upon graduation based on mathematics competency. *Indones. J. Electr. Eng. Comput. Sci.* **2017**, *5*, 684–690.
20. Ivanova, V.; Zlatanov, B. Implementation of fuzzy functions aimed at fairer grading of students' tests. *Educ. Sci.* **2019**, *9*, 214. [CrossRef]
21. Depren, S.K. Prediction of students' science achievement: An application of multivariate adaptive regression splines and regression trees. *J. Balt. Sci. Educ.* **2018**, *17*, 887–903. [CrossRef]
22. Shahiri, A.M.; Husain, W.; Rashid, N.A. A review on predicting student's performance using data mining techniques. *Proced. Comp. Sci.* **2015**, *72*, 414–422. [CrossRef]
23. Dutt, A.; Ismail, M.A.; Herawan, T. A systematic review on educational data mining. *IEEE Access* **2017**, *5*, 15991–16005. [CrossRef]
24. Alyahyan, E.; Düştegör, D. Predicting academic success in higher education: Literature review and best practices. *Int. J. Educ. Technol. High. Educ.* **2020**, *17*, 3. [CrossRef]
25. Pintelas, P.; Livieris, I.E. Special issue on ensemble learning and applications. *Algorithms* **2020**, *13*, 140. [CrossRef]
26. Breiman, L.; Friedman, J.H.; Olshen, R.A.; Stone, C.J. *Classification and Regression Trees*; Chapman and Hall/CRC: Boca Raton, FL, USA, 1984.
27. Steinberg, D. CART: Classification and regression trees. In *The Top Ten Algorithms in Data Mining*; Wu, X., Kumar, V., Eds.; Chapman and Hall/CRC: Boca Raton, FL, USA, 2009; pp. 179–202.
28. Izenman, A.J. *Modern Multivariate Statistical Techniques. Regression, Classification, and Manifold Learning*; Springer: New York, NY, USA, 2008.
29. Apté, C.; Weiss, S. Data mining with decision trees and decision rules. *Future Gener. Comput. Syst.* **1997**, *13*, 197–210. [CrossRef]
30. Salford Predictive Modeler. Available online: https://www.minitab.com/en-us/products/spm/ (accessed on 17 November 2020).
31. Wen, S.; Buyukada, M.; Evrendilek, F.; Liu, J. Uncertainty and sensitivity analyses of co-combustion/pyrolysis of textile dyeing sludge and incense sticks: Regression and machine-learning models. *Renew. Energy* **2019**, *151*, 463–474. [CrossRef]

32. Pradeepkumar, D.; Ravi, V. Forex rate prediction: A hybrid approach using chaos theory and multivariate adaptive regression splines. In *Proceedings of the 5th International Conference on Frontiers in Intelligent Computing: Theory and Applications*; Springer: Berlin/Heidelberg, Germany, 2017; Volume 515, pp. 219–227. [CrossRef]
33. Friedman, J.H. Multivariate adaptive regression splines (with discussion). *Ann. Stat.* **1991**, *19*, 1–141. [CrossRef]
34. Bliemel, F. Theil's forecast accuracy coefficient: A clarification. *J. Mark. Res.* **1973**, *10*, 444–446. [CrossRef]
35. Vandekerckhove, J.; Matzke, D.; Wagenmakers, E.J. Model comparison and the principle of parsimony. In *The Oxford Handbook of Computational and Mathematical Psychology*; Busemeyer, J.R., Wang, Z., Townsend, J.T., Eidelsm, A., Eds.; Oxford University Press: Oxford, UK, 2015; pp. 300–318.
36. Queiruga-Dios, A.; Sanchez, M.J.S.; Perez, J.J.B.; Martin-Vaquero, J.; Encinas, A.H.; Gocheva-Ilieva, S.; Demlova, M.; Rasteiro, D.D.; Caridade, C.; Gayoso-Martinez, V. Evaluating Engineering Competencies: A New Paradigm. In Proceedings of the Global Engineering Education Conference (EDUCON), Tenerife, Spain, 17–20 April 2018; IEEE: New York, NY, USA, 2018; pp. 2052–2055. [CrossRef]

Article

Improving the Accuracy of Dam Inflow Predictions Using a Long Short-Term Memory Network Coupled with Wavelet Transform and Predictor Selection

Trung Duc Tran, Vinh Ngoc Tran and Jongho Kim *

School of Civil and Environmental Engineering, University of Ulsan, Ulsan 44610, Korea; trungtd@mail.ulsan.ac.kr (T.D.T.); vinhtn@mail.ulsan.ac.kr (V.N.T.)
* Correspondence: kjongho@ulsan.ac.kr; Tel.: +82-052-259-2855

Citation: Tran, T.D.; Tran, V.N.; Kim, J. Improving the Accuracy of Dam Inflow Predictions Using a Long Short-Term Memory Network Coupled with Wavelet Transform and Predictor Selection. *Mathematics* 2021, 9, 551. https://doi.org/10.3390/math9050551

Academic Editor: Snezhana Gocheva-Ilieva

Received: 29 January 2021
Accepted: 2 March 2021
Published: 5 March 2021

Publisher's Note: MDPI stays neutral with regard to jurisdictional claims in published maps and institutional affiliations.

Copyright: © 2021 by the authors. Licensee MDPI, Basel, Switzerland. This article is an open access article distributed under the terms and conditions of the Creative Commons Attribution (CC BY) license (https://creativecommons.org/licenses/by/4.0/).

Abstract: Accurate and reliable dam inflow prediction models are essential for effective reservoir operation and management. This study presents a data-driven model that couples a long short-term memory (LSTM) network with robust input predictor selection, input reconstruction by wavelet transformation, and efficient hyper-parameter optimization by K-fold cross-validation and the random search. First, a robust analysis using a "correlation threshold" for partial autocorrelation and cross-correlation functions is proposed, and only variables greater than this threshold are selected as input predictors and their time lags. This analysis indicates that a model trained on a threshold of 0.4 returns the highest Nash–Sutcliffe efficiency value; as a result, six principal inputs are selected. Second, using additional subseries reconstructed by the wavelet transform improves predictability, particularly for flow peak. The peak error values of LSTM with the transform are approximately one-half to one-quarter the size of those without the transform. Third, for a *K* of 5 as determined by the Silhouette coefficients and the distortion score, the wavelet-transformed LSTMs require a larger number of hidden units, epochs, dropout, and batch size. This complex configuration is needed because the amount of inputs used by these LSTMs is five times greater than that of other models. Last, an evaluation of accuracy performance reveals that the model proposed in this study, called SWLSTM, provides superior predictions of the daily inflow of the Hwacheon dam in South Korea compared with three other LSTM models by 84%, 78%, and 65%. These results strengthen the potential of data-driven models for efficient and effective reservoir inflow predictions, and should help policy-makers and operators better manage their reservoir operations.

Keywords: dam inflow prediction; long short-term memory; wavelet transform; input predictor selection; hyper-parameter optimization

1. Introduction

Reservoirs and dams serve a variety of critical purposes, including flood mitigation, freshwater storage, irrigation, hydroelectric power, and ecological conservation. Substantial efforts have been made over the past century to develop optimal reservoir operating strategies. Proposing an optimal operating solution for a multipurpose reservoir is not straightforward because it can be affected by various factors, the most important of which is reservoir inflow estimates [1–3]. Accurate and reliable inflow forecasts are essential to effective reservoir operation [4,5]. Predictive models can be divided into process-based and data-driven varieties [6,7]. As process-based models use mathematical formulations based on physical principles, embracing state variables and fluxes that are theoretically observable and scalable [8], they provide a superior understanding of physical processes [9–12]. However, these models also require detailed data volumes and high computational costs for a study basin [13–16], and when simplified assumptions related to scaling problems are applied, their predictions can be accompanied by considerable amounts of uncertainty [17–21]. Empirical data-driven models are based on past observations. They are simple and easy to apply, do not directly consider the underlying physical processes, and rely solely on historical hydro-meteorological data, resulting in less input and fewer parameter data [22,23].

With advances in statistical and machine-learning techniques, data-driven models have attracted attention for their strong learning capabilities and suitability for modeling complex nonlinear processes [24–27]. Techniques such as artificial neural networks (ANNs), recurrent neural network (RNNs), support vector regression (SVR), genetic programming, multilayer perceptrons (MLPs), adaptive neuro-fuzzy inference systems, and long short-term memory (LSTM) can provide satisfactory outcomes in meteorological and hydrological prediction studies. One of the latest network architectures and a special type of RNN, LSTM overcomes the notorious problem of vanilla RNN, in which gradients disappear and explode when performing backpropagation over multiple timesteps [28,29]. This difficulty in long-range dependency learning can be addressed by using memory cells of an LSTM architecture with cell states that are maintained over time [30]. LSTM is therefore able to learn the nonlinearity of input variables with an arbitrary length and effectively capture long-term time dependencies. Prior studies have demonstrated that streamflow predictions of LSTM are more accurate than those of ANNs, RNNs, SVR, and MLP [31–34].

Proper input selection and data processing play a crucial role in achieving a high-performing data-driven model [27,35]. First, a thorough understanding of the underlying physical processes and available data are required to select the appropriate input. Inconsistent selection of inputs can lead to a loss of convergence in model-training or poor accuracy in model application [36]. Most previous studies (Table 1) have used a trial-and-error based on multiple scenarios of input combinations or ad-hoc selections for critical factors [31,33,34,37,38]. Statistical properties of the data series derived from principal component analysis and correlation analysis can help identify explanatory variables [38]. The cross-correlation function (CCF) and the partial autocorrelation function (PACF) are often used to analyze correlations between candidate inputs and output. Second, data-driven models may not be able to handle nonstationary data if preprocessing is not carried out properly [33]. Cleaning, normalization, transformation, and reduction of data can significantly improve accuracy [24,39,40]. A wavelet transform (WT) can effectively process nonstationary data by decomposing time series into multiple subseries of lower resolution, and extract nontrivial and potentially useful information from the original data [24]. It has been employed extensively to solve problems related to the diagnosis, classification, and forecasts of extreme weather events [33]. Although the individual effects of distinctive features in Table 1 have been demonstrated in several studies, investigating the combined impacts of these methods in dam inflow predictions has not yet been performed. Therefore, the best models and methodologies in this subject need to be revealed.

Table 1. Streamflow prediction studies using a data-driven model.

Study	Data-Driven Model	Predictor Selection	Data Processing	Hyper-Parameter Determination
Kratzert et al. [37]	LSTM	Ad-hoc	Normalization	Trial and error
Hu et al. [31]	ANN, LSTM	Ad-hoc	Normalization	Ad-hoc
Lee et al. [38]	AR, FFNN, RNN	Ad-hoc	Copula-based transformation	Trial and error
Ni et al. [33]	LSTM, CNN	Ad-hoc	WT	Ad-hoc
Xiang et al. [34]	LSTM	Ad-hoc	Moving average	Ad-hoc
Yang et al. [4]	ANN, RF, SVM	Visual *	Normalization	Trial and error
Ahmad and Hossain [41]	ANN	Visual *	Moving average	Trial and error
This study	LSTM	Optimal **	WT, Normalization	Optimization

Note: * Selecting visually from CCF and/or PACF plots; ** Selecting optimally from a "threshold" analysis.

The main objective of this study is to investigate the potential of the combined methods of LSTM, predictor selection, data processing, and hyper-parameter optimization, thereby developing a unified data-driven modeling framework that can produce accurate dam inflow predictions. Of specific interest are (1) the robust selection of principal inputs and their sequence lengths, (2) the transformation of input time series to better capture extremes, and (3) the efficient optimization of LSTM hyper-parameters. The rest of this study is organized as follows. Section 2 describes the methodologies of LSTM structure, input selection, wavelet transform, and hyper-parameter optimization. Section 3 provides the study area, dataset, and performance measures. Section 4 presents the experimental results and discussion, and a conclusion follows in Section 5.

2. Methodology

2.1. Long Short-Term Memory Network

Long short-term memory is a special kind of RNN that includes memory cells that are analogous to the states of physically based models [28]. An advantage of LSTM over an RNN is that LSTM can learn long-term dependencies between input and output features by resolving gradients that are exploding or vanishing [37]. The main difference between LSTM and RNN structures is that LSTM adds a cell state; four times more parameters should be trained because three gate functions are employed to calculate the cell and the hidden states. The internal structure of LSTM is sketched in Figure 1a.

A LSTM-based data-driven model is composed of repeating LSTM blocks, each of which contains three gates (forget gate f_t, input gate i_t, and output gate o_t) to determine which information is renewed, discarded, and outputted from the memory cell. Given the inputs $x_t = [x_{1,t}, x_{2,t}, \ldots, x_{N_{in},t}]$ at time t with the number of inputs N_{in}, cell state c_t (a long-term memory) and hidden state h_t (a short-term memory) at time t are computed using three gates and the cell state at a previous time step. A new state c_t can be controlled through a forget gate that can forget information from the past state c_{t-1} and an input gate that can accept new information from the cell update $\tilde{c}_t$. The output gate determines how much information from the cell state c_t flows into the new hidden state h_t. Mathematically,

$$c_t = f_t \times c_{t-1} + i_t \times \tilde{c}_t \tag{1}$$

$$h_t = o_t \times \tanh(c_t) \tag{2}$$

where the intermediate cell update $\tilde{c}_t$ and three gates are calculated for x_t and h_{t-1}:

$$\tilde{c}_t = \tanh(W_{\tilde{c}} x_t + U_{\tilde{c}} h_{t-1} + b_{\tilde{c}}) \tag{3}$$

$$f_t = \sigma\left(W_f x_t + U_f h_{t-1} + b_f\right) \tag{4}$$

$$i_t = \sigma(W_i x_t + U_i h_{t-1} + b_i) \tag{5}$$

$$o_t = \sigma(W_o x_t + U_o h_{t-1} + b_o) \tag{6}$$

where W, U, and b are learnable parameters specific to the three gates and determined through the training process. The activation functions of σ and $\tanh$ are the sigmoid and the hyperbolic tangent, respectively. At $t = 1$, the hidden and cell states are initialized as zero vectors [37]. LSTM models can also be built with more than one layer by stacking multiple layers on top of each other. The output of a stacked LSTM connects to a final "dense layer." A target output y_t can be computed from h_t in the dense layer:

$$y_t = W_d h_t + b_d \tag{7}$$

where W_d and b_d are learnable parameters known as the weight matrix and the bias term, respectively, of the dense layer. The total number of LSTM parameters is therefore 12 for each layer plus 2 for the dense layer. The shapes of these parameters can be expressed in a matrix (Table 2). At the beginning of training, the learnable parameters are initialized using an Xavier initialization and later optimized by the Adam algorithm, preferred in abundant studies [42].

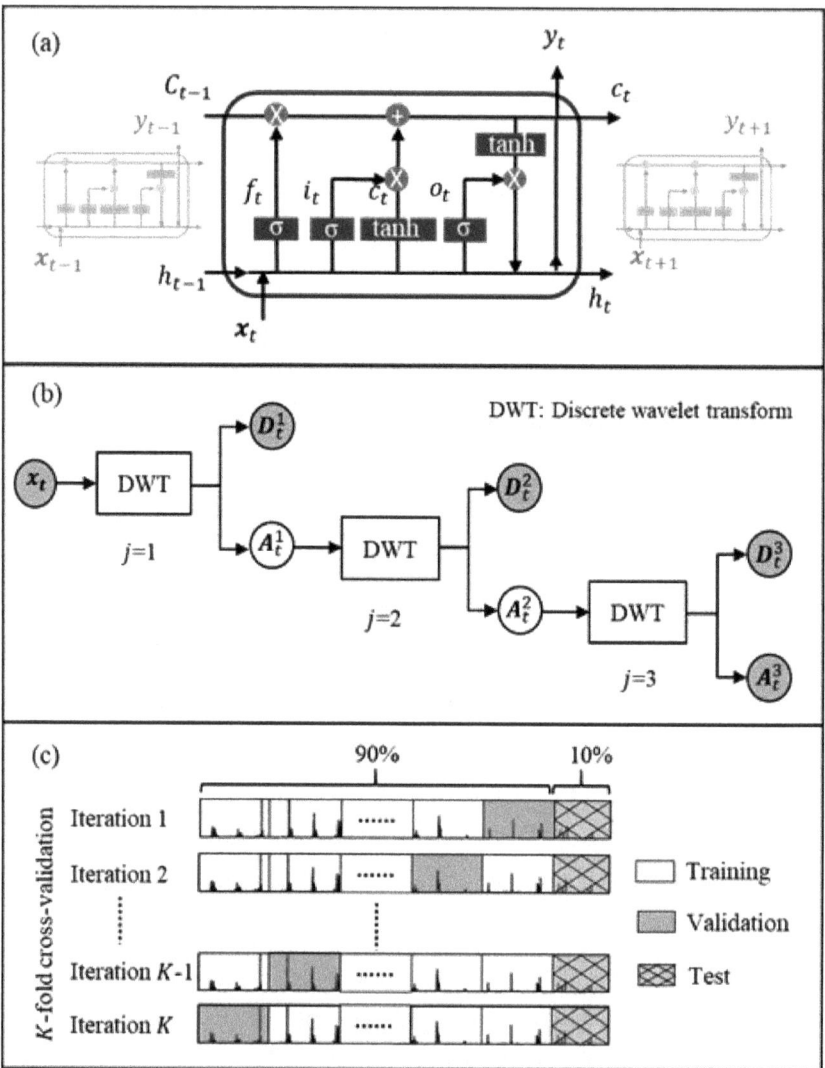

Figure 1. (a) The internal structure of long short-term memory (LSTM), where x_t and y_t denote input predictors and a target output; f, i, o stand for the forget, input, and output gate, respectively; c_t, $\tilde{c}_t$, and h_t denote cell state, cell update, and hidden state at time t, respectively. (b) Diagram of wavelet transform; the input time series x_t can be subdivided into multiple subseries down to the j-th level (i.e., D_t^j and A_t^j); D_t^j and A_t^j denote 'Detail' and 'Approximation' time series of the original x_t at the level j; the input predictors and their subseries (shown as gray circles) are all used as the input of the LSTM models. (c) Schematic overview of K-fold cross-validation. The training and validation dataset (corresponding to 90% of total dataset in this study) is randomly split into K folds. A fold (shown as gray color) is used to validate the LSTM trained for the other folds (white color) at each iteration.

Table 2. Learnable parameters of each layer of LSTM and their shapes.

Layer	Parameter	Shape
1	$W_{\tilde{c}}, W_f, W_i, W_o$ $U_{\tilde{c}}, U_f, U_i, U_o$ $b_{\tilde{c}}, b_f, b_i, b_o$	$[N_{hu}\ N_{in}]$ $[N_{hu}\ N_{hu}]$ $[N_{hu}]$
2	$W_{\tilde{c}}, W_f, W_i, W_o$ $U_{\tilde{c}}, U_f, U_i, U_o$ $b_{\tilde{c}}, b_f, b_i, b_o$	$[N_{hu}\ N_{hu}]$ $[N_{hu}\ N_{hu}]$ $[N_{hu}\ N_{hu}]$
$\vdots$	$\vdots$	$\vdots$
N_l	$W_{\tilde{c}}, W_f, W_i, W_o$ $U_{\tilde{c}}, U_f, U_i, U_o$ $b_{\tilde{c}}, b_f, b_i, b_o$	$[N_{hu}\ N_{hu}]$ $[N_{hu}\ N_{hu}]$ $[N_{hu}\ N_{hu}]$
Dense	W_d b_d	$[N_{hu}\ 1]$ $[1]$

2.2. Input Predictor Selection

Maintaining a high correlation between inputs and outputs can guarantee the predictability of data-driven models. Therefore, to arrive at the optimal combination of inputs that correlate closely with the output, statistical properties of the respective time series can be used. Specifically, the cross-correlation function (CCF) and the partial autocorrelation function (PACF) are used to determine the appropriate predictors and the number of lagged values.

The CCF measures the similarity of a time series (e.g., dam inflow, y) with its lagged versions (e.g., candidate input variables, $v = [v_1, v_2, \ldots, v_{N_v}]$ with N_v candidates):

$$\text{CCF}_k^{vy} = \frac{c_k^{vy}}{\sqrt{SD_v SD_y}} \tag{8}$$

where k is the lag time; SD_v and SD_y are the standard deviations of v and y, respectively; c_k^{vy}, which is the cross-covariance function of v and y, is defined as:

$$c_k^{vy} = \frac{1}{N_t - 1} \sum_{t=1}^{N_t - k} (y_t - \bar{y})(v_{t+k} - \bar{v}) \tag{9}$$

where t is the time step; $\bar{y}$ is the average of y; $\bar{v}$ is the average of v; N_t denotes the number of data points of time series.

The PACF measures the linear correlation between a time series (y_t) and a lagged version of itself (y_{t+k}) and can be defined as:

$$\text{PACF}_{k,k} = \frac{\rho_k - \sum_{j=1}^{k-1} \text{PACF}_{k-1,j} \times \rho_{k-j}}{1 - \sum_{j=1}^{k-1} \text{PACF}_{k-1,j} \times \rho_j} \tag{10}$$

where j denotes an index for lag k; ρ_k is an autocorrelation coefficient at lag k between y_t and y_{t+k}. At $k = 1$, $\text{PACF}_{1,1}$ is equal to ρ_1.

$$\rho_k = \frac{\sum_{t=1}^{N_t - k}(y_t - \bar{y})(y_{t+k} - \bar{y})}{\sum_{t=1}^{N_t}(y_t - \bar{y})^2} \tag{11}$$

Additionally, an approximate 95% confidence interval (CI) on the CCF and PACF [43] can be estimated by:

$$95\%\ \text{CI} = -\frac{1}{N_t} \pm \frac{2}{\sqrt{N_t}} \tag{12}$$

Based on the calculated CCF and PACF, modelers can determine the number of antecedent values that should be included in the input vector. Other variables that may not have a significant effect on model performance can be cut off from the input vector. Generally, the CCF can be used as a reference when selecting highly correlative input predictors, while the PACF can indicate an appropriate lag for the selected variables.

2.3. Wavelet Transform

A WT is a mathematical tool that decomposes one signal into several with lower resolution levels by controlling the scaling and shifting factors of a single wavelet—the mother wavelet function exists locally as a pattern. It offers time–frequency localization of a given time series and analyzes nonstationary elements such as breakdown points, discontinuities, and local minima and maxima [35]. Due to the nature of streamflow represented by discrete signals, a discrete wavelet transform (DWT) is usually preferred in hydrological applications [33]. A DWT is easier to implement compared with a continuous wavelet transform and has a shorter computational time [44]. However, a DWT is not inherently shift-invariant. If any new values are added to the end of a time series, certain values of the wavelet component can change. This means it cannot be applied to problems related to singularity detection, forecasting, and nonparametric regression [45]. To overcome this "boundary" problem, an à trous algorithm that uses redundant information attained from observation data has been suggested [46]. The decomposition formulas of an à trous algorithm are defined as [47]:

$$D^j_t = A^{j-1}_t - A^j_t \tag{13}$$

$$A^j_t = \sum_{l=0}^{L-1} g_l A^{j-1}_{t-2^{j-1}l \bmod N_t} \tag{14}$$

where D^j_t and A^j_t represent the jth-level wavelet (detail) and scaling (approximation) coefficients of the original time series at time t; g_l is a scaling filter with $g_l = g^{DWT}_l / \sqrt{2}$ where g^{DWT}_l is a scaling filter for DWT; L is the length of the scaling filter; l denotes an index for L; mod refers to the modulo operator. At $j = 0$, A^0_t is equal to the original time series of x_t. The latter can be obtained by the additive reconstruction:

$$x_t = \sum_{j=1}^{J} D^j_t + A^J_t \tag{15}$$

As depicted in Figure 1b, an original signal decomposes into D^1_t and A^1_t through the wavelet and scaling filters, and A^1_t further decomposes into D^2_t and A^2_t through the same process. This expansion is repeated until j reaches to the maximum level J. The number of decomposed subseries is J + 1. For example, if J = 3, the subseries would be $[D^1_t, D^2_t, D^3_t, A^3_t]$ for each original time series. The total number of subseries for N_{in} input variables is therefore $(J + 1) \times N_{in}$. The approximation A^j becomes increasingly rough as j increases. As data processing using a Daubechies 5 wavelet at level 3 has been preferred in studies of flow predictions [33,48–51], the discrete wavelet at level 3 (J = 3) was used in this study.

2.4. LSTM Hyper-Parameters

Configuring an LSTM network by adjusting hyper-parameters is a difficult task, but it can have a significant impact on the performance of data-driven models [37]. Additionally, the shape of the learnable parameters depends heavily on the number of inputs (N_{in}), the hyper-parameters of the number of hidden units (N_{hu}), and the number of layers (N_l), as shown in Table 2. As inappropriate values of N_{hu} and N_l can lead to unreliable LSTM models, close attention should be paid to their selection. If these values are too large, the learnable parameters that need to be trained will increase, the size of the training dataset will be large, and considerable training time will be required. Complicating matters further, too many hidden units can cause overfitting phenomena in data-driven models [52].

To compensate for this issue, a dropout technique is often used, as reducing the number of cells in the network during training can prevent overfitting. The number of cells can be adjusted from 0 to 1 depending on the dropout rate (N_D).

To train an LSTM network by estimating the learnable parameters W, U, and b, an objective function (or a loss function) for a given hyper-parameters must be evaluated. Here, the value of the loss function was computed from a subset (i.e., a mini-batch) of LSTM predictions and their corresponding observations; the learnable parameters during training were updated according to a given loss function at each iteration step. The number of iterations (N_{it}) was determined based on N_t, the mini-batch size (N_b), and the number of epochs (N_e) (i.e., $N_{it} = N_t/N_b \times N_e$). Neural networks using small batch sizes can achieve convergence with fewer epochs [53]. However, using an N_b that is too small can lead to a large number of iterations, which will take excessive time to compute them. For this study, Nash–Sutcliffe efficiency (NSE) was chosen as a loss function as it can build LSTM with greater prediction accuracy compared with other metrics, such as the mean square error [54].

The hyper-parameters associated with the configuration of this study consisted of mini-batch size (N_b), dropout rate (N_D), the number of hidden units (N_{hu}), the number of layers (N_l), and the number of epochs (N_e). When tuning the hyper-parameters, two popular approaches, grid search and random search, are often used [55]. In the first approach, the grid search can be considered exhaustive as it defines a search space as a grid of hyper-parameter values and evaluates grid position for all combinations of all hyper-parameter values. A random search defines a search space as a bounded domain of hyper-parameter values and chooses random combinations in that domain for evaluations. The latter approach can create a more reliable model with more combinations of hyper-parameters, particularly when large amounts of training are used [55–57]. A random search was therefore used to determine an appropriate set of hyper-parameters to optimize the LSTM network in this study.

We chose a special form of resampling procedure, K-ld cross-validation, to evaluate the LSTM model's performance with a limited dataset. First, the dataset was partitioned into equally (or nearly equally) K-sized folds or clusters. Subsequently, K iterations were performed for training and validation such that within each iteration, a different fold of the dataset was held out for validation (gray cells in Figure 1c) while the remaining K-1 folds were used for training (white cells in Figure 1c) [58]. A useful set of hyper-parameters can provide almost equally good validation values for an object function for each iteration. To determine an appropriate number of clusters (K), the average of Silhouette coefficients ($\bar{s}$) is commonly used [59]. For a given data point i in a cluster, the Silhouette coefficient $s(i)$ is defined as:

$$s(i) = \frac{b(i) - a(i)}{max\{a(i), b(i)\}} \tag{16}$$

where $a(i)$ is the average distance between point i and all other points in the same cluster, and $b(i)$ is the average distance between the point i and all points in the nearest cluster. It is advisable to choose a K that provides a high value of $\bar{s}$.

2.5. Summary of Modeling Framework

A brief overview of the methodology adopted for dam inflow prediction, hereafter called SWLSTM, follows the schematic in Figure 2:

(1) Collect the time series of both the target output y_t (i.e., dam inflow) and the candidate input predictors $v_t = [v_{1,t}, v_{2,t}, \ldots, v_{N_v,t}]$, $t = 1, \ldots, N_t$. Any inappropriate or missing values in the collected data should be reviewed carefully.
(2) Determine the explanatory "principal" variables $x_t = [x_{1,t}, x_{2,t}, \ldots, x_{N_{in},t}]$, $t = 1, \ldots, N_t$ among the candidate predictors, with an appropriate lag time, using the CCF and PACF.
(3) Decompose and reconstruct the selected input predictors into the wavelet-transformed subseries $\left[D^1_{1,t}, D^2_{1,t}, D^3_{1,t}, A^3_{1,t}, \ldots, D^1_{N_{in},t}, D^2_{N_{in},t}, D^3_{N_{in},t}, A^3_{N_{in},t}\right]$. These reconstructed data

are normalized to values between 0 and 1, and then split into one set for training and validation and another for testing. In this study, we set 90% of the total data length for training and validation and 10% for test.

(4) Determine the number of clusters for the K-fold cross-validation and then optimize five hyper-parameters by the random search over the training and validation set.

(5) Train and build the LSTM models, using the optimal values of hyper-parameters over the training and validation set.

(6) Assess and compare the performance of LSTM models in predicting dam inflow for the test dataset. The LSTM models chosen to demonstrate the effectiveness of SWLSTM presented here are (1) a regular LSTM without both the determination of principal lags and variables and the WT, (2) a "WLSTM," which is a regular LSTM coupled with a WT, and (3) a "SLSTM," which is similar to a regular LSTM but performs the input specification in the Step 2.

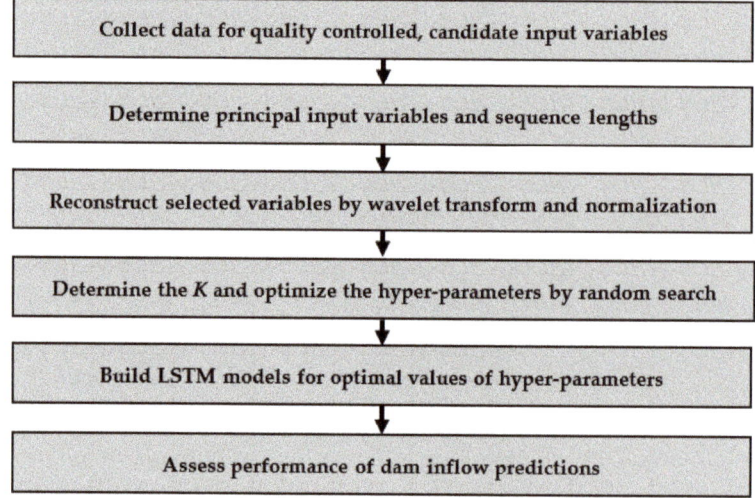

Figure 2. Flow chart of the methodologies adopted for predicting dam inflow.

2.6. Evaluation Metrics

Three evaluation metrics of NSE, mean absolute error (MAE), and peak error (PE) were used to qualitatively measure the performance of model accuracy. Each was computed over the test period as:

$$\text{NSE} = 1 - \frac{\sum_{t=1}^{N_t} \left(y_t - y_t^{obs}\right)^2}{\sum_{t=1}^{N_t} \left(y_t^{obs} - \overline{y^{obs}}\right)^2} \tag{17}$$

$$\text{MAE} = \frac{\sum_{t=1}^{N_t} |y_t - y_t^{obs}|}{N_t} \tag{18}$$

$$\text{PE} = \frac{|y_{max}^{obs} - y_{max}|}{y_{max}^{obs}} \times 100 \tag{19}$$

where y_t^{obs} denotes the observation of dam inflow at time t; $\overline{y^{obs}}$ is the average of y^{obs}; y_{max} and y_{max}^{obs} are the predicted and observed peak dam inflow, respectively.

2.7. Open Source Software

Our research relies on open source software with the programing language of Python 3.7 [60]. The libraries of Numpy [61], Pandas [62] and Scikit-learn [63] were used for managing and preprocessing the data. TensorFlow [64] and Keras [65] were utilized to implement LSTM. The hardware environment was configured with Intel(R) Xeon(R) Gold 6242 CPU at 2.80 GHz × 32 processors, and 376 GB of RAM.

3. Study Area and Dataset

The Hwacheon dam watershed in the central part of the Korean Peninsula (its latitude and longitude are 127°47′ E and 38°7′ N, respectively) was chosen as a case study (Figure 3). The watershed created by the dam covers 3901 km^2, approximately 80% of which is forest, and its elevation varies from close to 120 m at the dam site to 1600 m. The Hwacheon dam was designed as a multipurpose dam for generate electricity, prevent floods, and store water. Its power generation capacity is 326 GWh and its total storage capacity is approximately 1018 Mm^3, making it a relatively large dam for South Korea. The Peace dam, located upstream of the Hwacheon dam, was built to prevent flooding and prepare for North Korean military (flood) attacks, and is normally operated as a dry-water dam. A daily dataset for 5844 days from 1 January 2004, to 31 December 2019 was collected and is described in Table 3. Wherein, the first 5260 days (90% of the total data) were used for training and validation, while the rest was left for testing the trained models. The data collected includes inflow to the Hwacheon dam (Q_{in}), dam outflow from Peace dam (Q_o), and meteorological data such as precipitation (Pr), temperature (Ta), humidity (H), wind speed (Ws), and pressure (Pre). Any inappropriate (negative) or missing values in the collected data were replaced with those interpolated linearly. The amount of such an inappropriate data is however less than 0.01%. Spatially averaged precipitation was computed using Thiessen polygons for six rain gauges in Table 3 and Figure 3.

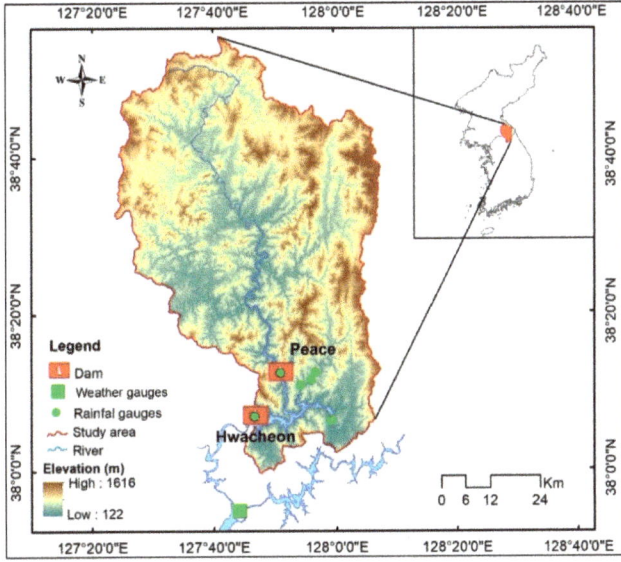

Figure 3. Geological location and topographic characteristics of the 'Hwacheon' dam watershed located in the central part of the Korean Peninsula.

Table 3. Information about the data used for predicting the inflow of the Hwacheon dam.

Variable *	Station Name	Station ID	Longitude	Latitude	Source
Qin (m^3/s)	Hwacheon dam	1010310	127°46′60″	38° 7′0″	
Qo (m^3/s)	Peace dam	1009710	127°50′55″	38°12′43″	Water Resources Management Information System (http://www.wamis.go.kr:8081/ENG/, accessed on 1st January 2020)
Pr (mm)	Hwacheongunchung	10094010	127°50′54″	38°12′34″	
	Bangsanchogyo	10104030	127°56′35″	38°12′36″	
	Hwacheondam	10104050	127°46′38″	38° 7′2″	
	Geumakri	10104060	127°55′52″	38°11′36″	
	Suibcheon	10104170	127°54′5″	38°10′59″	
	Yanggu Seocheon	10104171	127°59′3″	38° 6′28″	
Ta (°C)	Chuncheon	101	127°44′8.51″	37°54′59.27″	Automated Surface Observing System (https://data.kma.go.kr/cmmn/main.do, accessed on 1st January 2020)
H (%)					
Ws (m/s)					
Pre (hPa)					

* Qin: inflow to Hwacheon dam; Qo: outflow from Peace dam; Pr: precipitation; Ta: temperature; H: humidity; Ws: wind speed; Pre: pressure.

4. Results and Discussion

4.1. Determining Principal Input Predictors and Their Sequence Lengths

To construct an appropriate input combination for the LSTM model, the principal variables and sequence lengths were determined using the statistical properties of the candidate variables. Figure 4 provides the statistical correlations between the seven candidate input variables (i.e., Qin, Qo, Pr, Ta, H, Ws, and Pre) and the target variable (Qin). Figure 4a shows the CCF between the seven candidate variables with a time lag of zero, indicating that Qo and Pr were strongly correlated with Qin (their CCFs were approximately 0.63 and 0.48, respectively); Ta and H had a relatively weaker correlation with Qin; and Ws and Pre had a negative correlation with Qin. The correlations for other combinations of the remaining variables were all less than 0.2, with the exception of the correlation between H and Ta. Regarding the correlations between Qin and the seven candidate variables at different time lags from 0 to 10, Figure 4b shows that Qin had a strong autocorrelation up to a 1-day lag (the PACF is approximately 0.8), and the correlation became significantly lower when the time lag was greater than 1 day. The CCF values for Qo and Pr, which had a strong correlation with Qin, became smaller as the time lag increased, while the values for the remaining four variables were not influenced by the magnitude of the time delay (see Figure 4c–h).

Based on the PACF and CCF analyses, a final set of input variables and sequence lengths (time lags) that had a high correlation with the target output Qin were selected. However, choosing only the input variables and the lags that have a close correlation with the target variable posed some challenges. In previous research, such a selection was typically based on user decisions made through trial and error, and no specific rules or criteria were used to determine the which key inputs were optimal [5,37,66]. We proposed a robust analysis using a "correlation threshold" for the PCAF and CCF values, and only variables greater than this threshold were used as input predictors and their time lags to construct and train a model. If the correlation threshold was small (e.g., 0.026, the upper bound of the 95% CI from Equation (12)), most of the variables and sequence lengths could be adopted to predict the target variable. Conversely, if the threshold was large (e.g., 0.8), the SLSTM model was constructed using only a limited number of predictors that were highly correlated with the target variable. Figure 5 depicts the performance of the SLSTM against the correlation threshold as a loss function (NSE). A model trained on a threshold of 0.4 produced the highest NSE value of 0.66, which was considered optimal. The model performed the poorest with a small threshold of 0.026, which means that a large number of inputs (up to 44 in this study) that were not highly correlated with the target variable led to overfitting and less-accurate outcomes. However, if using a high threshold (greater than 0.6), the number of input predictors may be limited (only 1 in this study) and not be sufficient to describe the target variable. By selecting the principal variables and their corresponding sequence lengths based on the optimal threshold of 0.4, Qin_{t-1}, Qo_{t-1}, Qo_{t-2}, Qo_{t-3}, Pr_{t-1}, and Pr_{t-2} became the inputs to predict the inflow of Hwacheon dam.

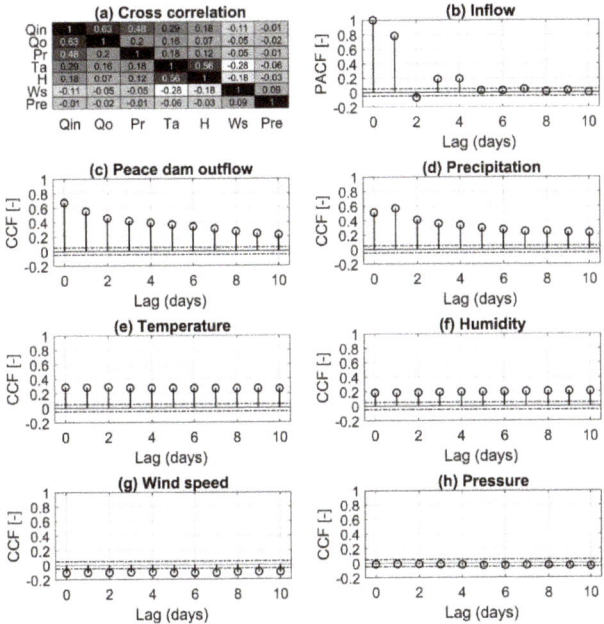

Figure 4. (**a**) Cross-correlation functions (CCF) between inflow to Hwacheon dam (Qin) and candidate input variables of outflow from Peace dam (Qo), precipitation (Pr), temperature (Ta), humidity (H), wind speed (Ws), and pressure (Pre); (**b**) partial autocorrelation functions (PACF) with time lags for Qin; (**c**–**h**) cross-correlation functions between Qin and the candidate variables at different lags. The black dot lines denote 95% confidence interval computed from Equation (12).

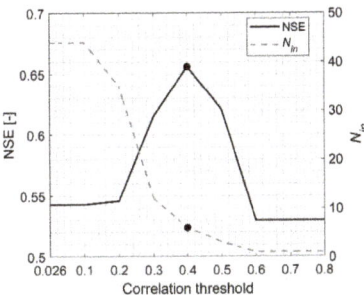

Figure 5. The effects of a correlation threshold for the PCAF and CCF values on the NSE of SLSTM over the validating period below (see the black line). The dashed line on the right axis denotes the number of inputs (N_{in}) including principal predictors and their time lags for each threshold. The black dots indicate the optimal threshold value. The first 4680 days (80%) data and the next 580 days (10%) are employed for training and validation, respectively. The hyper-parameters suggested by Kratzert, Klotz [37] are employed.

4.2. Decomposing Input Time Series by a Wavelet Transform

Three levels of DWT decomposition were performed on the seven candidate input variables for WLSTM and the six selected inputs defined above for SWLSTM, extracting four subseries for each. Figure 6 shows the original and transformed time series of the three principal variables of Qin, Qo, and Pr. The degree of fluctuations in the "Detail" time series is smoother and has a lower frequency at higher decomposition levels. The "Approximation" A^3 has a rougher and slower gradual trend than the original time series x.

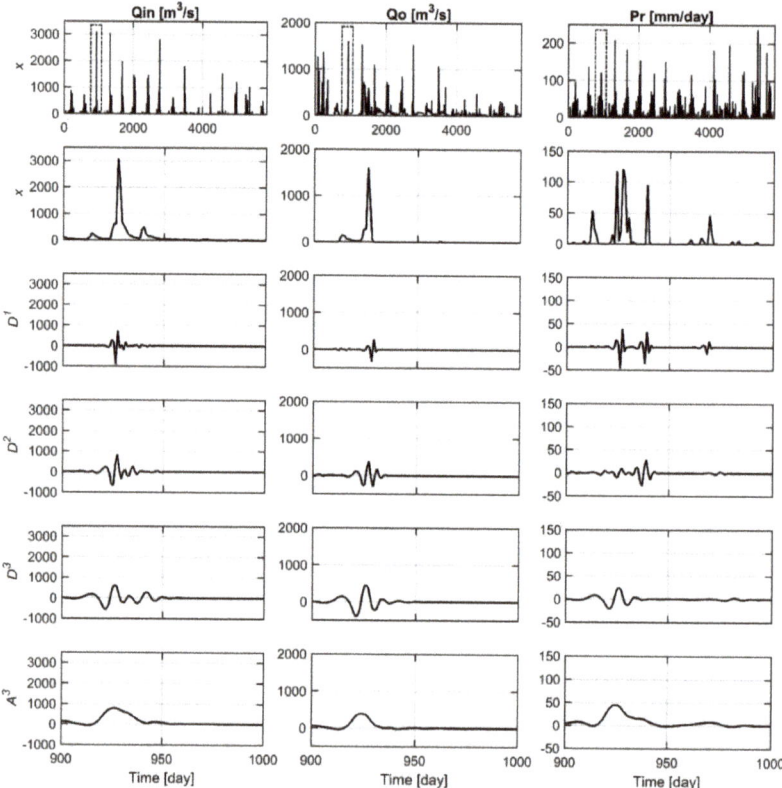

Figure 6. Decomposed time series for the three principal input variables (Qin, Qo, and Pr) after wavelet transform. x denotes the original time series; D^1, D^2, and D^3 denote the 'Detail' time series at the levels of 1, 2, and 3, respectively; A^3 refers to the 'Approximation' time series among the decompositions of x at the level 3. The subplots from the second row to the end row are zoomed in for the peak of the subplots of the first row (a period of 900–1000).

Data processing by the three levels of decomposition created an additional time series that is five times the number of the original time series (i.e., one original time series plus four decomposed time series). Eventually, the inputs used to train the LSTM models presented in this study and predict dam inflow were (1) for LSTM, all seven candidate variables at t-1 timestep (i.e., Qin_{t-1}, Qo_{t-1}, Pr_{t-1}, Ta_{t-1}, H_{t-1}, Ws_{t-1}, and Pre_{t-1}); (2) for SLSTM, six variables selected from the above "correlation threshold" analysis (i.e., Qin_{t-1}, Qo_{t-1}, Qo_{t-2}, Qo_{t-3}, Pr_{t-1}, and Pr_{t-2}); (3) for WLSTM, in addition to the seven candidate variables, subtime series on each by WT (for a total of 35 inputs); (4) for SWLSTM, a total of 30 variables made by WT on the six variables. For more details, see Table 4.

Table 4. Summary of input and output variables used for training four data-driven models (LSTM, SLSTM, WSLTM, and SWLSTM).

Model	x	y
LSTM	$Qin_{t-1}, Qo_{t-1}, Pr_{t-1}, Ta_{t-1}, H_{t-1}, Ws_{t-1}, Pre_{t-1}$	
SLSTM	$Qin_{t-1}, Qo_{t-1}, Qo_{t-2}, Qo_{t-3}, Pr_{t-1}, Pr_{t-2}$	
WLSTM	$Qin_{t-1}, D^1_{Qin,\ t-1}, D^2_{Qin,\ t-1}, D^3_{Qin,\ t-1}, A^3_{Qin,\ t-1},$ $Qo_{t-1}, D^1_{Qo,\ t-1}, D^2_{Qo,\ t-1}, D^3_{Qo,\ t-1}, A^3_{Qo,\ t-1},$ $Pr_{t-1}, D^1_{Pr,\ t-1}, D^2_{Pr,\ t-1}, D^3_{Pr,\ t-1}, A^3_{Pr,t-1},$ $Ta_{t-1}, D^1_{Ta,\ t-1}, D^2_{Ta,t-1}, D^3_{Ta,t-1}, A^3_{Ta,t-1},$ $H_{t-1}, D^1_{H,t-1}, D^2_{H,t-1}, D^3_{H,\ t-1}, A^3_{H,t-1},$ $Ws_{t-1}, D^1_{Ws,t-1}, D^2_{Ws,t-1}, D^3_{Ws,t-1}, A^3_{Ws,t-1},$ $Pre_{t-1}, D^1_{Pre,t-1}, D^2_{Pre,t-1}, D^3_{Pre,t-1}, A^3_{Pre,t-1}$	Qin_t
SWLSTM	$Qin_{t-1}, D^1_{Qin,\ t-1}, D^2_{Qin,\ t-1}, D^3_{Qin,\ t-1}, A^3_{Qin,\ t-1},$ $Qo_{t-1}, D^1_{Qo,\ t-1}, D^2_{Qo,\ t-1}, D^3_{Qo,\ t-1}, A^3_{Qo,\ t-1},$ $Qo_{t-2}, D^1_{Qo,\ t-2}, D^2_{Qo,\ t-2}, D^3_{Qo,\ t-2}, A^3_{Qo,\ t-2},$ $Qo_{t-3}, D^1_{Qo,\ t-3}, D^2_{Qo,\ t-3}, D^3_{Qo,\ t-3}, A^3_{Qo,\ t-3},$ $Pr_{t-1}, D^1_{Pr,\ t-1}, D^2_{Pr,\ t-1}, D^3_{Pr,\ t-1}, A^3_{Pr,t-1},$ $Pr_{t-2}, D^1_{Pr,\ t-2}, D^2_{Pr,\ t-2}, D^3_{Pr,\ t-2}, A^3_{Pr,t-2}$	

4.3. Optimizing the Hyper-Parameters

Choosing an appropriate set of hyper-parameters significantly affected model performance. To investigate the effects of the five hyper-parameters on the value of the loss function (i.e., NSE), their configurations were set as listed in Table 5, with controlled values suggested by Kratzert, Klotz [37]. As shown in Figure 7, change in NSE values was negligible (neither increasing nor decreasing) as the value of each hyper-parameter increased. SWLSTM consistently provided the largest NSE, with values in the range of 0.7–0.8, while LSTM provided the smallest NSE range of 0.5–0.6. These results indicate in part that SWLSTM outperformed the other models.

Table 5. The configurations of the hyper-parameters to investigate the effects of each hyper-parameter on the model performance in Figure 7. The controlled value for each hyper-parameter was borrowed from Kratzert, Klotz [37].

	The Number of Layers	The Number of Hidden Units	Dropout Rate	Batch Size	The Number of Epochs
Figure 7a	1, 2, 3, 4, 5, 6, 7	100	0.1	512	200
Figure 7b	1	10, 50, 100, 150, 200, 250, 300, 350, 400	0.1	512	200
Figure 7c	1	20	0.1, 0.2, 0.3, 0.4, 0.5, 0.6, 0.7, 0.8, 0.9	512	200
Figure 7d	1	20	0.1	2, 4, 8, 16, 32, 64, 128, 256, 512	200
Figure 7e	1	20	0.1	512	10, 50, 100, 200, 250, 300, 350, 400, 450, 500

In this study, an optimal hyper-parameter set for the four data-driven models was determined from K-fold cross-validation and the random search. First, an appropriate number of clusters for K-fold cross-validation was chosen based on the average Silhouette coefficients, $\bar{s}$, and the "distortion score" of the elbow method (Figure 8). In general, K values can be selected that provide a high value of $\bar{s}$ and those corresponding to the "elbow" of the distortion score curve. As K increased, the $\bar{s}$ value tended to decrease overall, so K values from 2 to 5 could be chosen. The gray dashed line in Figure 8 shows an elbow with a K near 5–7. A K of 5 was therefore selected to optimize the hyper-parameters of the four data-driven models in this study.

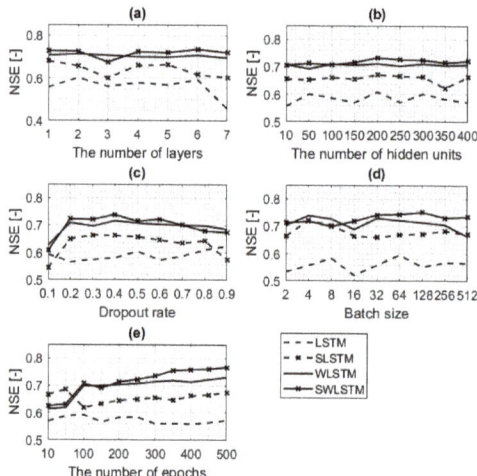

Figure 7. The effects of hyper-parameters, (**a**) the number of layers, (**b**) the number of hidden units, (**c**) dropout rate, (**d**) batch size and (**e**) the number of epochs on NSE, computed over the validating period using the four data-driven models (LSTM, SLSTM, WLSTM, and SWLSTM). The first 4680 days (80%) data and the next 580 days (10%) are employed for training and validation, respectively. Hyper-parameters used for each subplot are illustrated in Table 5.

For a *K* of 5 and a confined range specified in the second column of Table 6, a set of hyper-parameters for the four models was found using the random search method (see Table 6 for the optimal hyper-parameter set). As a result, two LSTM layers were required for all four models, while the values of other hyper-parameters varied. Specifically, WLSTM and SWLSTM required a larger number of hidden units, epochs, dropout, and batch size compared with the other two models. Such a complex configuration was required because the amount of inputs used by WLSTM and SWLSTM was five times more than in other models. The CPU time required for this process is 2–6 min, much more than LSTM training that takes about 50 s and other processes (LSTM prediction, wavelet transformation, and normalization) that only take a few seconds.

Table 6. The prior ranges of the hyper-parameters used in the Random search (see the second column) and determined optimal values of the hyper-parameters for four data-driven models by random search and K-fold cross-validation with *K* of 5 (see the rest columns).

Hyper-Parameters	Range	LSTM	SLSTM	WLSTM	SWLSTM
The number of layers	[1–7]	2	2	2	2
The number of hidden units	[10–400]	100	150	200	200
The number of epochs	[10–500]	250	250	250	250
Dropout rate	[0.1–0.9]	0.3	0.5	0.5	0.6
Batch size	[2–512]	8	8	32	32

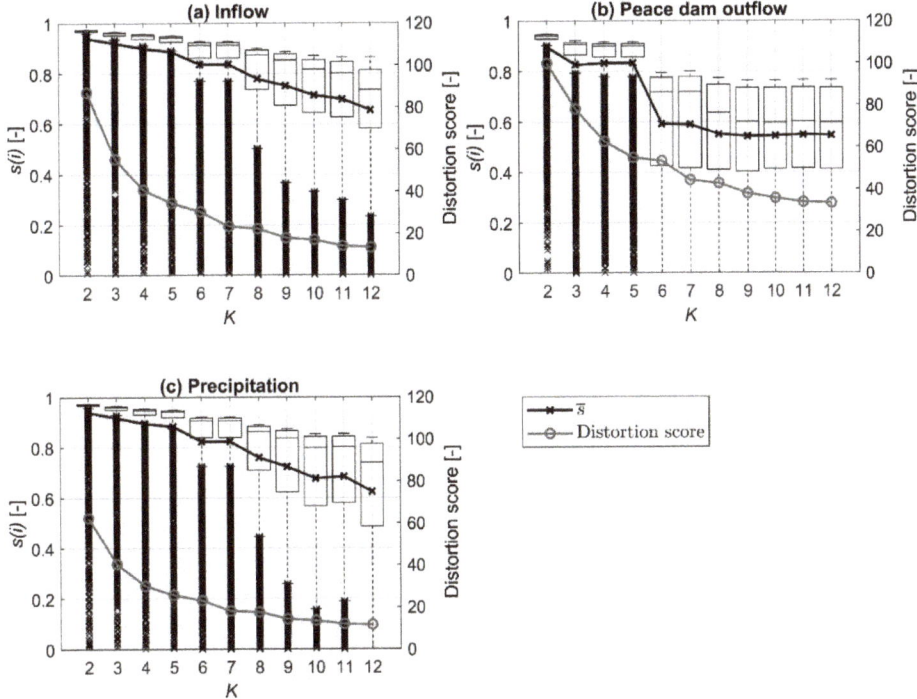

Figure 8. Silhouette coefficient $s(i)$ versus the number of clusters (K) used in K-fold cross-validation. Boxplots are drawn from 5260 s values (i.e., $i = 5260$) corresponding to the entire length of training and validation dataset for the three principal variables (i.e., Qin, Qo, and Pr). The boxplots demonstrate the median (central mark), the 25th and 75th percentiles (the edges of the box), and the maximum and minimum (the upper and lower whiskers) except for outliers (cross symbols). The black-cross line ($\bar{s}$) demonstrates the average values of $s(i)$. The gray lines on the right axis represent the 'distortion score' computed in the elbow method.

4.4. Predicting Dam Inflow with Trained LSTMs

LSTM models trained using optimal hyper-parameters were applied to the test dataset to predict inflow at the Hwacheon dam. Comparing the hydrographs with observations, the overall variation and magnitude of predicted inflow using SWLSTM agreed more closely with observations than did the results produced by other models (Figure 9). Quantitatively, SWLSTM had an R^2 of 0.96 in a 1:1 comparison between prediction and observation, which outperformed the 0.77, 0.92, and 0.92 values produced by LSTM, SLSTM, and WL-STM, respectively. The results produced by NSE, MAE, and PE confirm that the predictions from SWLSTM were closest to observations. In particular, compared with LSTM, which has an NSE of 0.65 and a PE of 29.1%, both metrics were significantly improved to an NSE of 0.89 and a PE of 7.7% (see Figure 9 for specific values).

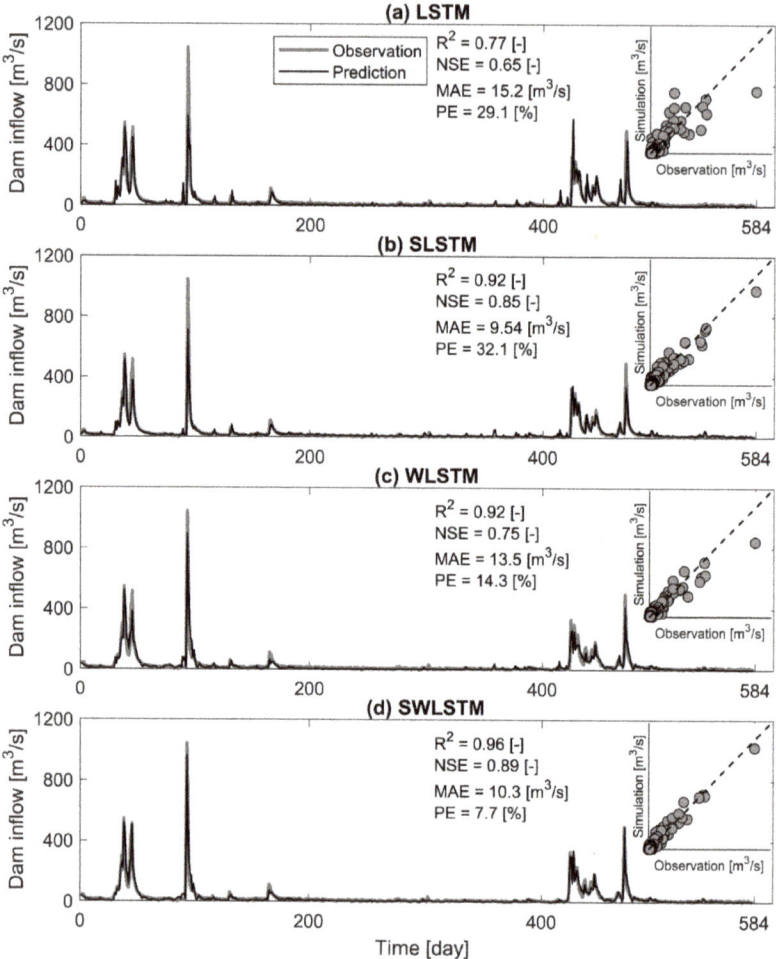

Figure 9. The comparisons of the dam inflow predicted by four models of (a) LSTM, (b) SLSTM, (c) WLSTM, and (d) SWLSTM with observations for the reserved 'test' dataset. The subplots demonstrate 1:1 comparisons between observations and predictions of dam inflow at each timestep. The optimal hyper-parameters specified in Table 6 and the 5-folds (K = 5) are employed.

To examine the superiority of SWLSTM proposed in this study compared with the other models, a relative "difference" metric (Δ) in Equation (20) was introduced:

$$\Delta = \frac{|Metric_A - Metric_{ideal}| - |Metric_B - Metric_{ideal}|}{|Metric_A - Metric_{ideal}|} \times 100 \qquad (20)$$

where $Metric_A$ and $Metric_B$ are the values of a metric for two models A and B, $Metric_{ideal}$ represents the ideal (perfect) value of the metrics of R^2, NSE, MAE, and PE (1, 1, 0, and 0, respectively). The positive (or negative) values of Δ indicate that the prediction results of model B are more (or less) accurate than those computed by model A. Table 7 shows that the accuracy performance of SWLSTM was superior to that of LSTM, SLSTM, and WLSTM.

Table 7. Accuracy improvements of SWLSTM to three other models of LSTM, SLSTM, and WLSTM for dam inflow predictions for the test dataset. A relative "difference" metric (Δ_{Metric}) in Equation (20) is computed for four evaluation metrics (R2, NSE, MAE, and PE). The positive (or negative) values of Δ_{Metric} indicate that the prediction results of SWLSTM are more (or less) accurate than those computed by other models. The optimal hyper-parameters specified in Table 6 and the 5-folds (K = 5) are employed.

Relative "Difference" Metric (%)	SWLSTM vs. LSTM	SWLSTM vs. SLSTM	SWLSTM vs. WLSTM
Δ_{R^2}	82.5	50	49.5
Δ_{NSE}	68.4	27.7	53.4
Δ_{MAE}	29.8	−8.4	25.5
Δ_{PE}	75	77.3	48.8

Based on these results, we concluded that the selection of appropriate input predictors and time lags helped create a more reliable data-driven model (see comparisons of SLSTM vs. LSTM and SWLSTM vs. WLSTM). In general, all factors related to weather and hydrology as well as the past histories of each variable, can affect dam inflow predictions, but using too much information that is not highly correlated (Figure 4) creates an overfitting model. As the number of input data increases, the noise for that input will also increase, and it is difficult to accurately estimate weights (or learnable parameters) for each input. Additionally, the historical period that affects the future inflow varies depending on the predictor. For example, Qin is closely related to itself a day ago, while for Qo and Pr, data from 3 and 2 days ago are also important.

It is interesting to note that the use of WT improved the flood peak predictability of data-driven models. That is, the PE values of WLSTM and SWLSTM were approximately one-half and one-quarter the size of those from LSTM and SLSTM, respectively. Additionally, near the flood peak in the test dataset, the bias values for both models using WT were much smaller than those for the non-WT models (Figure 10). With the aid of WT, five times as much data were used, and they can be reconstructed with various levels of decomposition. To train extreme data such as flood peaks, including separate time series such as WT appeared to be effective. If only one original data point was used for training, abnormally extreme high-frequency data may be considered noisy rather than critical information to be learned, which may fail to recognize, learn, and predict these events.

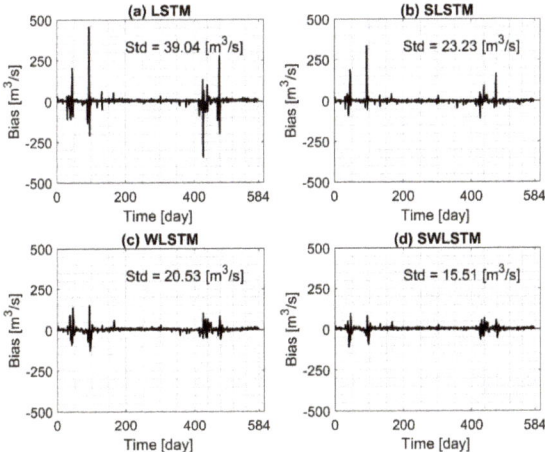

Figure 10. Bias between observation and prediction of dam inflow for four models in test dataset; Std is the standard deviation. The optimal hyper-parameters specified in Table 6 and the 5-folds (K = 5) are employed.

4.5. Feasibility to Multimodal, Multitask, and Bidirectional Learning

Recently, multimodal, multitask, and bidirectional learning has received great interest [33,34,66,67], and it is worth discussing the feasibility of hydrological time series predictions (e.g., dam inflow, runoff, or flood predictions). First, both multimodal learning with multiple inputs and multitask learning with multiple outputs are related to high-dimensional problems. In hydrological time series predictions, there are three reasons for increasing the dimension, namely, a case where the number of input or output variables is large, a case where the sequence length or lead time of each variable is long, and a case where the values of the variables vary spatially. In the latter case, the number of dimensions can increase significantly up to $O(10^2$ to $10^6)$ while it is not very large, $O(10^0$ to $10^1)$ in the first two cases. In this study, multimodal learning for the first two cases was performed with a maximum of 30 dimensions. As a future study, it will be necessary to review the learning ability for ultra-high-dimensional problems that can take into account the spatial heterogeneity of variables.

The second is to examine the applicability of bidirectional learning in predicting hydrologic time series. Several studies have mentioned the superiority of bidirectional learning [68], which combines information from both the past and the future at the same time. However, this is limited to predictions of language models or hindcasting problems in which future data exist. It is challenging to apply it to hydrological forecasting cases because future information about weather variables (e.g., precipitation) and human (e.g., dam) operation is unknown at the present time.

5. Conclusions

In this study, data-driven models based on an LSTM network were built to predict daily inflow at the Hwacheon dam in South Korea. Three important aspects were considered to improve the accuracy of dam inflow predictions in an integrated fashion: (1) principal input predictors and their time lags were determined from a robust analysis of the statistical properties of the data series; (2) the original time series was converted to multiscale subseries by a WT; (3) hyper-parameters of all models were efficiently optimized through K-fold cross-validation and the random search. The effectiveness of SWLSTM, a model trained to consider these aspects, was compared with LSTM (trained without input selection and data transformation), SLSTM (trained with input selection only), and WLSTM (trained with data transformation only). The primary findings of this study are presented in the following paragraphs.

First, seven candidate input variables (i.e., inflow to the Hwacheon dam, outflow from Peace dam, precipitation, temperature, humidity, wind speed, and pressure) were initially chosen to investigate the correlation properties for the target output, Qin. Based on PACF and CCF analyses, we selected a final set of input variables and their sequence lengths (time lags). However, how to choose only the input variables and the lags that are closely correlated with the target variable remains an open question. In this study, a robust analysis using a correlation threshold for the PCAF and CCF values was proposed, and only variables greater than this threshold were selected as input predictors and their time lags. As shown in Figure 5, a model trained on a threshold of 0.4 produced the highest NSE value. Eliminating variables that have a low correlation with Qin helped prevent divergence and restrict overfitting in the learning model. Conclusively, Qin_{t-1}, Qo_{t-1}, Qo_{t-2}, Qo_{t-3}, Pr_{t-1}, and Pr_{t-2} become the principal inputs to predict the inflow of the dam. The effectiveness of such an input specification was validated because the models using it (SLSTM and SWLSTM) provided exceptionally accurate predictions compared with the unused models (i.e., LSTM and WLSTM).

Second, using additional data series reconstructed by a WT improved predictability, particularly for flow peak (see the comparisons of WLSTM vs. LSTM and SWLSTM vs. SLSTM). The PE values of WLSTM and SWLSTM were approximately one-half and one-quarter the size of those produced by LSTM and SLSTM, respectively. For training extreme data such as flow peaks, including separate time series by WT can be effective. If only one original data point is used for training, abnormally extreme high-frequency data may be considered noisy rather than critical information to be learned, and the system may fail to recognize, learn, and predict these events.

Third, for a K of 5 as determined by the Silhouette coefficients and the distortion score (Figure 8), a set of hyper-parameters for the four models was found using a random search (Table 6). Both WLSTM and SWLSTM require a larger number of hidden units, epochs, dropout, and batch size compared with the other two models. The need for this complex configuration is clear because the amount of inputs used by WLSTM and SWLSTM was five times greater than that of the other models.

Last, accuracy performance investigated by various evaluation metrics revealed that SWLSTM is superior to LSTM, SLSTM, and WLSTM by 84%, 78%, and 65%, respectively. When the SWLSTM framework in this study is coupled with the procedures of a WT and the input specifications, overall and peak accuracy of time-dependent flow prediction improved. Ultimately, accurate forecasts of inflow will help policy makers and operators better manage their reservoir operations and tasks.

Author Contributions: Conceptualization, T.D.T. and J.K.; methodology, T.D.T., V.N.T. and J.K.; formal analysis, T.D.T., V.N.T. and J.K.; writing—original draft preparation, T.D.T.; writing—review and editing, T.D.T., V.N.T. and J.K.; visualization, T.D.T.; funding acquisition, J.K. All authors have read and agreed to the published version of the manuscript.

Funding: This work was supported by KOREA HYDRO and NUCLEAR POWER CO., LTD (No.2019-Tech-11) and the National Research Foundation of Korea (NRF) grant funded by the Korea government (MSIT) (NRF-2019R1C1C1004833).

Institutional Review Board Statement: Not applicable.

Informed Consent Statement: Not applicable.

Data Availability Statement: Data was contained within the article.

Conflicts of Interest: The authors declare no conflict of interest.

References

1. Jothiprakash, V.; Magar, R.B. Multi-time-step ahead daily and hourly intermittent reservoir inflow prediction by artificial intelligent techniques using lumped and distributed data. *J. Hydrol.* **2012**, *450–451*, 293–307. [CrossRef]
2. El-Shafie, A.; Taha, M.R.; Noureldin, A. A neuro-fuzzy model for inflow forecasting of the Nile river at Aswan high dam. *Water Resour. Manag.* **2006**, *21*, 533–556. [CrossRef]
3. Seo, Y.; Kim, S.; Kisi, O.; Singh, V.P. Daily water level forecasting using wavelet decomposition and artificial intelligence techniques. *J. Hydrol.* **2015**, *520*, 224–243. [CrossRef]
4. Yang, T.; Asanjan, A.A.; Welles, E.; Gao, X.; Sorooshian, S.; Liu, X. Developing reservoir monthly inflow forecasts using artificial intelligence and climate phenomenon information. *Water Resour. Res.* **2017**, *53*, 2786–2812. [CrossRef]
5. Zemzami, M.; Benaabidate, L. Improvement of artificial neural networks to predict daily streamflow in a semi-arid area. *Hydrol. Sci. J.* **2016**. [CrossRef]
6. He, Z.H.; Tian, F.Q.; Gupta, H.V.; Hu, H.C.; Hu, H.P. Diagnostic calibration of a hydrological model in a mountain area by hydrograph partitioning. *Hydrol. Earth Syst. Sci.* **2015**, *19*, 1807–1826. [CrossRef]
7. Chen, Y.; Zhou, H.; Zhang, H.; Du, G.; Zhou, J. Urban flood risk warning under rapid urbanization. *Env. Res* **2015**, *139*, 3–10. [CrossRef] [PubMed]
8. Fatichi, S.; Vivoni, E.R.; Ogden, F.L.; Ivanov, V.Y.; Mirus, B.; Gochis, D.; Downer, C.W.; Camporese, M.; Davison, J.H.; Ebel, B.; et al. An overview of current applications, challenges, and future trends in distributed process-based models in hydrology. *J. Hydrol.* **2016**, *537*, 45–60. [CrossRef]
9. Kim, J.; Ivanov, V.Y. On the nonuniqueness of sediment yield at the catchment scale: The effects of soil antecedent conditions and surface shield. *Water Resour. Res.* **2014**, *50*, 1025–1045. [CrossRef]
10. Kim, J.; Ivanov, V.Y.; Katopodes, N.D. Hydraulic resistance to overland flow on surfaces with partially submerged vegetation. *Water Resour. Res.* **2012**, *48*. [CrossRef]
11. Kim, J.; Dwelle, M.C.; Kampf, S.K.; Fatichi, S.; Ivanov, V.Y. On the non-uniqueness of the hydro-geomorphic responses in a zero-order catchment with respect to soil moisture. *Adv. Water Resour.* **2016**, *92*, 73–89. [CrossRef]
12. Warnock, A.; Kim, J.; Ivanov, V.; Katopodes, N.D. Self-Adaptive Kinematic-Dynamic Model for Overland Flow. *J. Hydraul. Eng.* **2014**, *140*, 169–181. [CrossRef]

13. Tran, V.N.; Dwelle, M.C.; Sargsyan, K.; Ivanov, V.Y.; Kim, J. A Novel Modeling Framework for Computationally Efficient and Accurate Real-Time Ensemble Flood Forecasting With Uncertainty Quantification. *Water Resour. Res.* **2020**, *56*. [CrossRef]
14. Tran, V.N.; Kim, J. Quantification of predictive uncertainty with a metamodel: Toward more efficient hydrologic simulations. *Stoch. Environ. Res. Risk Assess.* **2019**, *33*, 1453–1476. [CrossRef]
15. Clark, M.P.; Bierkens, M.F.P.; Samaniego, L.; Woods, R.A.; Uijlenhoet, R.; Bennett, K.E.; Pauwels, V.R.N.; Cai, X.; Wood, A.W.; Peters-Lidard, C.D. The evolution of process-based hydrologic models: Historical challenges and the collective quest for physical realism. *Hydrol. Earth Syst. Sci.* **2017**, *21*, 3427–3440. [CrossRef] [PubMed]
16. Tran, V.N.; Kim, J. Toward an Efficient Uncertainty Quantification of Streamflow Predictions Using Sparse Polynomial Chaos Expansion. *Water* **2021**, *13*, 203. [CrossRef]
17. Kim, J.; Ivanov, V.Y. A holistic, multi-scale dynamic downscaling framework for climate impact assessments and challenges of addressing finer-scale watershed dynamics. *J. Hydrol.* **2015**, *522*, 645–660. [CrossRef]
18. Kim, J.; Lee, J.; Kim, D.; Kang, B. The role of rainfall spatial variability in estimating areal reduction factors. *J. Hydrol.* **2019**, *568*, 416–426. [CrossRef]
19. Dwelle, M.C.; Kim, J.; Sargsyan, K.; Ivanov, V.Y. Streamflow, stomata, and soil pits: Sources of inference for complex models with fast, robust uncertainty quantification. *Adv. Water Resour.* **2019**, *125*, 13–31. [CrossRef]
20. Kim, J.; Ivanov, V.Y.; Fatichi, S. Environmental stochasticity controls soil erosion variability. *Sci. Rep.* **2016**, *6*, 22065. [CrossRef]
21. Kim, J.; Ivanov, V.Y.; Fatichi, S. Soil erosion assessment-Mind the gap. *Geophys. Res. Lett.* **2016**, *43*, 12446–12456. [CrossRef]
22. Kratzert, F.; Klotz, D.; Herrnegger, M.; Sampson, A.K.; Hochreiter, S.; Nearing, G.S. Toward Improved Predictions in Ungauged Basins: Exploiting the Power of Machine Learning. *Water Resour. Res.* **2019**, *55*, 11344–11354. [CrossRef]
23. Marcais, J.; de Dreuzy, J.R. Prospective Interest of Deep Learning for Hydrological Inference. *Ground Water* **2017**, *55*, 688–692. [CrossRef]
24. Nourani, V.; Hosseini Baghanam, A.; Adamowski, J.; Kisi, O. Applications of hybrid wavelet–Artificial Intelligence models in hydrology: A review. *J. Hydrol.* **2014**, *514*, 358–377. [CrossRef]
25. Aksoy, H.; Dahamsheh, A. Markov chain-incorporated and synthetic data-supported conditional artificial neural network models for forecasting monthly precipitation in arid regions. *J. Hydrol.* **2018**, *562*, 758–779. [CrossRef]
26. Yaseen, Z.M.; El-shafie, A.; Jaafar, O.; Afan, H.A.; Sayl, K.N. Artificial intelligence based models for stream-flow forecasting: 2000–2015. *J. Hydrol.* **2015**, *530*, 829–844. [CrossRef]
27. Shen, C. A Transdisciplinary Review of Deep Learning Research and Its Relevance for Water Resources Scientists. *Water Resour. Res.* **2018**, *54*, 8558–8593. [CrossRef]
28. Hochreiter, S.; Schmidhuber, J. Long short-term memory. *J. Neural Comput.* **1997**, *9*, 1735–1780. [CrossRef] [PubMed]
29. Bengio, Y.; Simard, P.; Frasconi, P. Learning long-term dependencies with gradient descent is difficult. *IEEE Trans. Neural Netw.* **1994**, *5*, 157–166. [CrossRef]
30. Greff, K.; Srivastava, R.K.; Koutnik, J.; Steunebrink, B.R.; Schmidhuber, J. LSTM: A Search Space Odyssey. *IEEE Trans. Neural Netw. Learn. Syst.* **2017**, *28*, 2222–2232. [CrossRef] [PubMed]
31. Hu, C.; Wu, Q.; Li, H.; Jian, S.; Li, N.; Lou, Z. Deep Learning with a Long Short-Term Memory Networks Approach for Rainfall-Runoff Simulation. *Water* **2018**, *10*, 1543. [CrossRef]
32. Le, H.; Lee, J. Application of Long Short-Term Memory (LSTM) Neural Network for Flood Forecasting. *Water* **2019**, *11*, 1387. [CrossRef]
33. Ni, L.; Wang, D.; Singh, V.P.; Wu, J.; Wang, Y.; Tao, Y.; Zhang, J. Streamflow and rainfall forecasting by two long short-term memory-based models. *J. Hydrol.* **2020**, *583*, 124296. [CrossRef]
34. Xiang, Z.; Yan, J.; Demir, I. A Rainfall-Runoff Model With LSTM-Based Sequence-to-Sequence Learning. *Water Resour. Res.* **2020**, *56*. [CrossRef]
35. Adamowski, J.; Sun, K. Development of a coupled wavelet transform and neural network method for flow forecasting of non-perennial rivers in semi-arid watersheds. *J. Hydrol.* **2010**, *390*, 85–91. [CrossRef]
36. Bowden, G.J.; Maier, H.R.; Dandy, G.C. Input determination for neural network models in water resources applications. Part 2. Case study: Forecasting salinity in a river. *J. Hydrol.* **2005**, *301*, 93–107. [CrossRef]
37. Kratzert, F.; Klotz, D.; Brenner, C.; Schulz, K.; Herrnegger, M. Rainfall–runoff modelling using Long Short-Term Memory (LSTM) networks. *Hydrol. Earth Syst. Sci.* **2018**, *22*, 6005–6022. [CrossRef]
38. Lee, T.; Shin, J.-Y.; Kim, J.-S.; Singh, V.P. Stochastic simulation on reproducing long-term memory of hydroclimatological variables using deep learning model. *J. Hydrol.* **2020**, *582*, 124540. [CrossRef]
39. Ravansalar, M.; Rajaee, T.; Kisi, O. Wavelet-linear genetic programming: A new approach for modeling monthly streamflow. *J. Hydrol.* **2017**, *549*, 461–475. [CrossRef]
40. Zhang, H.; Singh, V.P.; Wang, B.; Yu, Y. CEREF: A hybrid data-driven model for forecasting annual streamflow from a socio-hydrological system. *J. Hydrol.* **2016**, *540*, 246–256. [CrossRef]
41. Ahmad, S.K.; Hossain, F. A generic data-driven technique for forecasting of reservoir inflow: Application for hydropower maximization. *Environ. Model. Softw.* **2019**, *119*, 147–165. [CrossRef]
42. Kingma, D.P.; Ba, J. Adam: A Method for Stochastic Optimization. *arXiv* **2014**, arXiv:1412.6980.
43. Box, G.E.P.; Jenkins, G.M.; Reinsel, G.C. *Time Series Analysis: Forecasting and Control*, 4th ed.; Wiley: Hoboken, NJ, USA, 2008. [CrossRef]

44. Belayneh, A.; Adamowski, J.; Khalil, B.; Quilty, J. Coupling machine learning methods with wavelet transforms and the bootstrap and boosting ensemble approaches for drought prediction. *Atmos. Res.* **2016**, *172–173*, 37–47. [CrossRef]
45. Maheswaran, R.; Khosa, R. Comparative study of different wavelets for hydrologic forecasting. *Comput. Geosci.* **2012**, *46*, 284–295. [CrossRef]
46. Shensa, M.J. The discrete wavelet transform: Wedding the a trous and Mallat algorithms. *IEEE Trans. Signal Process.* **1992**, *40*, 2464–2482. [CrossRef]
47. Quilty, J.; Adamowski, J. Addressing the incorrect usage of wavelet-based hydrological and water resources forecasting models for real-world applications with best practices and a new forecasting framework. *J. Hydrol.* **2018**, *563*, 336–353. [CrossRef]
48. Budu, K. Comparison of Wavelet-Based ANN and Regression Models for Reservoir Inflow Forecasting. *J. Hydrol. Eng.* **2014**, *19*, 1385–1400. [CrossRef]
49. Nayak, P.C.; Venkatesh, B.; Krishna, B.; Jain, S.K. Rainfall-runoff modeling using conceptual, data driven, and wavelet based computing approach. *J. Hydrol.* **2013**, *493*, 57–67. [CrossRef]
50. Nourani, V.; Komasi, M.; Mano, A. A Multivariate ANN-Wavelet Approach for Rainfall–Runoff Modeling. *Water Resour. Manag.* **2009**, *23*, 2877–2894. [CrossRef]
51. Venkata Ramana, R.; Krishna, B.; Kumar, S.R.; Pandey, N.G. Monthly Rainfall Prediction Using Wavelet Neural Network Analysis. *Water Resour. Manag.* **2013**, *27*, 3697–3711. [CrossRef]
52. Hinton, G.E.; Srivastava, N.; Krizhevsky, A.; Sutskever, I.; Salakhutdinov, R.R. Improving neural networks by preventing co-adaptation of feature detectors. *arXiv* **2012**, arXiv:1207.0580.
53. Das, D.; Avancha, S.; Mudigere, D.; Vaidynathan, K.; Sridharan, S.; Kalamkar, D.; Kaul, B.; Dubey, P. Distributed Deep Learning Using Synchronous Stochastic Gradient Descent. *arXiv* **2016**, arXiv:1602.06709.
54. Kratzert, F.; Klotz, D.; Shalev, G.; Klambauer, G.; Hochreiter, S.; Nearing, G. Towards learning universal, regional, and local hydrological behaviors via machine learning applied to large-sample datasets. *Hydrol. Earth Syst. Sci.* **2019**, *23*, 5089–5110. [CrossRef]
55. Bergstra, J.; Bengio, Y. Random Search for Hyper-Parameter Optimization. *J. Mach. Learn. Res.* **2012**, *13*, 281–305.
56. Mantovani, R.G.; Rossi, A.L.D.; Vanschoren, J.; Bischl, B.; De Carvalho, A.C. Effectiveness of Random Search in SVM hyper-parameter tuning. In Proceedings of the 2015 International Joint Conference on Neural Networks (IJCNN), Killarney, Ireland, 12–17 July 2015; pp. 1–8.
57. Wu, L.; Perin, G.; Picek, S. I Choose You: Automated Hyperparameter Tuning for Deep Learning-based Side-channel Analysis. *Cryptol. Eprint Arch.* **2020**, *2020*, 1293.
58. Refaeilzadeh, P.; Tang, L.; Liu, H. Cross-Validation. In *Encyclopedia of Database Systems*; Liu, L., ÖZsu, M.T., Eds.; Springer: Boston, MA, USA, 2009; pp. 532–538. [CrossRef]
59. Rousseeuw, P.J. Silhouettes: A graphical aid to the interpretation and validation of cluster analysis. *J. Comput. Appl. Math.* **1987**, *20*, 53–65. [CrossRef]
60. Rossum, G. *Python Reference Manual*; CWI (Centre for Mathematics and Computer Science): Amsterdam, The Netherlands, 1995.
61. Van der Walt, S.; Colbert, S.C.; Varoquaux, G. The NumPy Array: A Structure for Efficient Numerical Computation. *Comput. Sci. Eng.* **2011**, *13*, 22–30. [CrossRef]
62. McKinney, W. Data Structures for Statistical Computing in Python. In Proceedings of the 9th Python in Science Conference, Austin, TX, USA, 28 June–3 July 2010.
63. Pedregosa, F.; Varoquaux, G.; Gramfort, A.; Michel, V.; Thirion, B.; Grisel, O.; Blondel, M.; Müller, A.; Nothman, J.; Louppe, G.; et al. Scikit-learn: Machine Learning in Python. *arXiv* **2012**, arXiv:1201.0490.
64. Abadi, M.; Agarwal, A.; Barham, P.; Brevdo, E.; Chen, Z.; Citro, C.; Corrado, G.S.; Davis, A.; Dean, J.; Devin, M.; et al. TensorFlow: Large-Scale Machine Learning on Heterogeneous Distributed Systems. *arXiv* **2016**, arXiv:1603.04467.
65. Chollet, F.O. Keras: Deep Learning Library for Theano and Tensorflow. 2015. Available online: https://github.com/fchollet/keras (accessed on 19 March 2019).
66. Dobrescu, A.; Giuffrida, M.V.; Tsaftaris, S.A. Doing More With Less: A Multitask Deep Learning Approach in Plant Phenotyping. *Front. Plant Sci.* **2020**, *11*, 141. [CrossRef]
67. Aceto, G.; Ciuonzo, D.; Montieri, A.; Pescapé, A. DISTILLER: Encrypted traffic classification via multimodal multitask deep learning. *J. Netw. Comput. Appl.* **2021**, 102985. [CrossRef]
68. Du, S.; Li, T.; Yang, Y.; Horng, S.-J. Multivariate time series forecasting via attention-based encoder–decoder framework. *Neurocomputing* **2020**, *388*, 269–279. [CrossRef]

Article

Motor Imagery Classification Based on a Recurrent-Convolutional Architecture to Control a Hexapod Robot

Tat'y Mwata-Velu [1], Jose Ruiz-Pinales [1], Horacio Rostro-Gonzalez [1], Mario Alberto Ibarra-Manzano [1], Jorge Mario Cruz-Duarte [2] and Juan Gabriel Avina-Cervantes [1,*]

1. Telematics and Digital Signal Processing Research Groups (CAs), Department of Electronics Engineering, University of Guanajuato, Salamanca 36885, Mexico; t.mwatavelu@ugto.mx (T.M.-V.); pinales@ugto.mx (J.R.-P.); hrostrog@ugto.mx (H.R.-G.); ibarram@ugto.mx (M.A.I.-M.)
2. Tecnológico de Monterrey, Monterrey 64849, Mexico; jorge.cruz@tec.mx
* Correspondence: avina@ugto.mx; Tel.: +52-4646479940 (ext. 2400)

Citation: Mwata-Velu, T.; Ruiz-Pinales, J.; Rostro-Gonzalez, H.; Ibarra-Manzano, M.A.; Cruz-Duarte, J.M.; Avina-Cervantes, J.G. Motor Imagery Classification Based on a Recurrent-Convolutional Architecture to Control a Hexapod Robot. *Mathematics* **2021**, *9*, 606. https://doi.org/10.3390/math9060606

Academic Editors: Vince Grolmusz and Snezhana Gocheva-Ilieva

Received: 10 February 2021
Accepted: 8 March 2021
Published: 12 March 2021

Publisher's Note: MDPI stays neutral with regard to jurisdictional claims in published maps and institutional affiliations.

Copyright: © 2021 by the authors. Licensee MDPI, Basel, Switzerland. This article is an open access article distributed under the terms and conditions of the Creative Commons Attribution (CC BY) license (https://creativecommons.org/licenses/by/4.0/).

Abstract: Advances in the field of Brain-Computer Interfaces (BCIs) aim, among other applications, to improve the movement capacities of people suffering from the loss of motor skills. The main challenge in this area is to achieve real-time and accurate bio-signal processing for pattern recognition, especially in Motor Imagery (MI). The significant interaction between brain signals and controllable machines requires instantaneous brain data decoding. In this study, an embedded BCI system based on fist MI signals is developed. It uses an Emotiv EPOC+ Brainwear®, an Altera SoCKit® development board, and a hexapod robot for testing locomotion imagery commands. The system is tested to detect the imagined movements of closing and opening the left and right hand to control the robot locomotion. Electroencephalogram (EEG) signals associated with the motion tasks are sensed on the human sensorimotor cortex. Next, the SoCKit processes the data to identify the commands allowing the controlled robot locomotion. The classification of MI-EEG signals from the F3, F4, FC5, and FC6 sensors is performed using a hybrid architecture of Convolutional Neural Networks (CNNs) and Long Short-Term Memory (LSTM) networks. This method takes advantage of the deep learning recognition model to develop a real-time embedded BCI system, where signal processing must be seamless and precise. The proposed method is evaluated using k-fold cross-validation on both created and public Scientific-Data datasets. Our dataset is comprised of 2400 trials obtained from four test subjects, lasting three seconds of closing and opening fist movement imagination. The recognition tasks reach 84.69% and 79.2% accuracy using our data and a state-of-the-art dataset, respectively. Numerical results support that the motor imagery EEG signals can be successfully applied in BCI systems to control mobile robots and related applications such as intelligent vehicles.

Keywords: brain-computer interface; EEG motor imagery; CNN-LSTM architectures; real-time motion imagery recognition

1. Introduction

In the last decade, various practical applications have been developed using the electrical signals generated in the brain through Brain-Computer Interfaces (BCIs) [1,2]. In particular, the Electroencephalograms (EEGs) are examples of such signals used currently to control specific devices such as service robots, motorized wheelchairs, drones, and several other human support machines. EEG signals have also found many applications in health sciences to detect mental diseases. For instance, a multi-modal machine learning approach was used to detect dementia integrating EEG-engineered features [3]. Some neurodegenerative disorders of the brain such as Alzheimer's disease have been studied using EEG signals to improve disease detection [4]. Early detection of schizophrenia risks in children between nine and 13 years was developed using EEG patterns classified by traditional machine learning algorithms [5].

Alternatively, using BCIs in specialized applications helps people with motor disabilities move or communicate with the surrounding environment [6,7]. Azmy et al. experimented with a BCI based on brain activation in the scalp using EEG signals for a robot's remote control [8]. Likewise, Palankar et al. tested a BCI to command a mobile robot using electronic systems from different suppliers. Indeed, integrating dissimilar technologies into BCIs has led to new research applications [9]. Among the mobile platforms, hexapod robots have excellent adaptability to terrain due to their zoomorphic structure. They present a quasi-controlled balance of their entire structure [10]. Hence, the experimental BCI contributions based on hexapod robots are still growing [11].

Several BCI architectures use a central computer as the primary signal processing unit [12,13]. Nevertheless, recent works have tested more versatile processing units [14], for instance using a dedicated Raspberry Pi board [15,16], implementing a Finite Impulse Response (FIR) filter on an Advanced RISC Machine (ARM) Linux embedded environment [17], or processing biological signals on embedded systems and the Internet of Things (IoT) [18]. Recently, Belwafi et al. used an Altera Stratix® IV Field-Programmable Gate Array (FPGA) to develop an embedded BCI system for home devices' control [19]. This trend is currently supported to meet the actual speed requirements of powerful signal processors using embedded systems [19,20].

The main challenges in the embedded BCI based on EEG signals are the classification accuracy, processing speed, low cost development, low power consumption, and hardware reconfiguration capabilities [21–23]. In the BCI area, the Convolutional Neural Networks (CNNs) and Long Short-Term Memory (LSTM) networks were recently implemented to classify EEG signals [24,25]. Jun Yang et al. studied the dynamic correlation between the temporal and spatial representation of Motor Imagery (MI)-EEG signals applied to BCI recognition using a CNN-LSTM model and the wavelet transform [24]. Torres et al. implemented a CNN on an FPGA applying an open-source project to convert the trained network model to an executable binary library for FPGA acceleration [26].

This study concerns developing a real-time embedded BCI system based on MI-EEG signals' recognition, using the Emotiv EPOC+ headset and an FPGA Altera Cyclone® V System on a Chip (SoC) card and applied in the locomotion of a hexapod robot. MI-EEG signals are wirelessly sent to the SoCKit platform to generate the corresponding commands, allowing robot displacement. For this project, signal processing modules were integrated into the SoCKit board. MI-EEG signals for closing-opening the right and left fist were captured by the F3, F4, FC5, and FC6 sensors and classified by a CNN-LSTM. Such hand movements were defined as tasks for motion commands sent directly to the hexapod. Under controlled conditions, a local dataset was built by carefully training four subjects, two males and two females aged from 23 to 36 years old, respectively. For comparison, a public database (Scientific-Data) was also used to validate the proposed embedded-BCI system by selecting a hand MI task subset related to the opening and closing of the left or right fist [27]. Finally, k-fold cross-validation was used to evaluate the commands' recognition rate used to control the hexapod locomotion.

2. Materials and Methods

The proposed methodology focuses on developing a BCI system by interconnecting an Emotiv EPOC+ headset, an Altera SoCKit Cyclone V SoC board, and a hexapod robot for validation. For this purpose, we created a database using basic MI movements (forward, backward, and stop) with four test subjects. The recognition process was carried out using a CNN-LSTM architecture. The operational system transforms the continuous MI-EEG signals into command instructions to control a hexapod robot's locomotion.

2.1. Proposed Framework

The EEG signal acquisition system was chosen following pragmatic criteria such as market accessibility, portability, resolution, sampling rate, compatibility, and scalability. An Emotiv EPOC+ headset consists of sixteen electrodes to be placed on the scalp according to the 10–20 international system of EEG electrode placement. Two of these sixteen electrodes are references, and fourteen are reserved for real-time capturing. On the other hand, the SoCKit module was used to process the MI signals to extract reliable robot commands. It is extensively described in Section 2.2.

In this study, robot perception and control were conceived to operate in real time. Simultaneously, data were stored on the SoCKit. The zoomorphic robot used has two degrees of freedom in each leg. In total, twelve servomotors quickly achieve static and kinematic stability. Figure 1 shows the proposed method's flowchart using an Emotiv EPOC+ for capturing EEG signals and a CNN-LSTM architecture implemented on the SoCKit to control a robot.

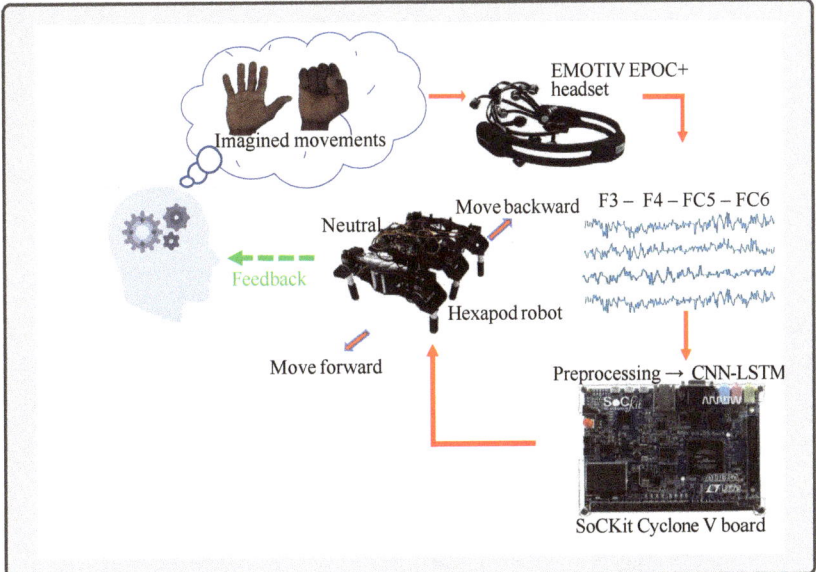

Figure 1. Flowchart of the proposed methodology. The EEG signals for the imagined closing-opening of the right and left fist are captured by the F3, F4, FC5, and FC6 sensors and sent to the SoCKit platform. The CNNs process the EEG data and extract feature sequences. A recurrent neural network followed by a dense layer classifies these feature sequences into robot locomotion commands.

2.2. A SoCKit Module Configuration

The SoCKit Cyclone V FPGA card powered by an ARM Cortex® A9 Hard Processor System (HPS) was used to implement the EEG signal processing algorithms and the classifier. Figure 2 shows the basic SoCKit functional blocks.

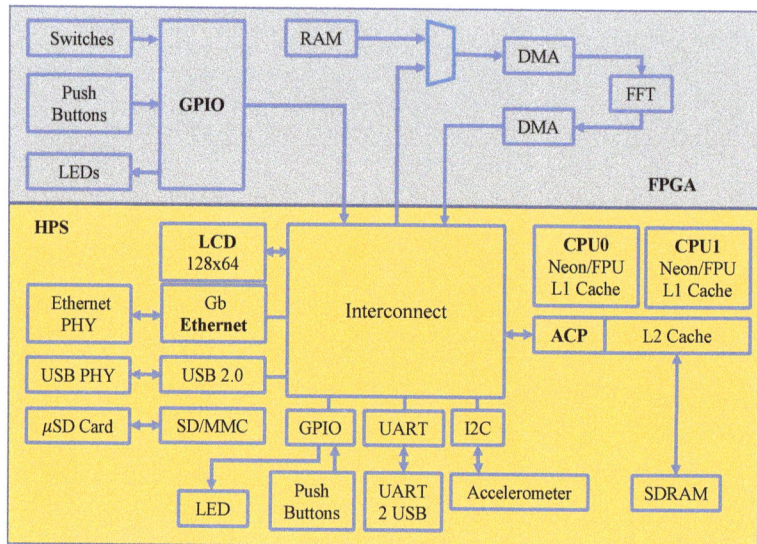

Figure 2. The SoCKit functional diagram consists of two main parts: the basic FPGA module and a Hard Processor System (HPS). Main modules include the Floating Point Unit (FPU), Accelerator Coherency Port (ACP), Secure Digital MultiMedia Card (SD/MMC), and Direct Memory Access (DMA).

The algorithms' implementation was coded and tested under the Xillybus for SoCKit Linux distribution (Xillinux) (https://www.terasic.com.tw/wiki/images/e/ef/Xillybus_getting_started_sockit.pdf, accessed on 7 February 2021) based on Ubuntu 12.04 LTS.

Communications were established between the processor and the FPGA core by configuring the Xillybus Intellectual Properties (IPs) Core, as showed in Figure 3.

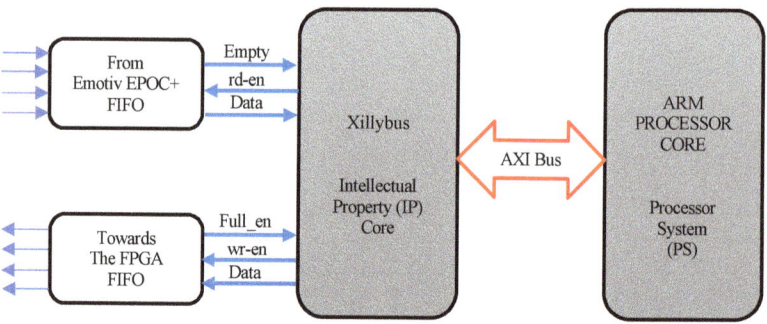

Figure 3. The Xillybus IPs Core was used as a data transport mechanism and configured to interconnect the processor core with the FPGA. The primary control signals include the write enable (wr-en), read enable (rd-en), and FIFO full enable (Full-en). Adapted from Xillibus Ltd.

Emotiv EPOC+ data writing (rd-en) is enabled as First-In, First-Out (FIFO) when it is empty. After reading the data, the Xillybus communicates with the processor core using the Advanced eXtensible Interface (AXI) bus, generating Direct Memory Access (DMA) requests on the central CPU bus. Simultaneously, the low-level FIFO (FPGA) is released (the full-en signal is low), and Xillybus carries the data from the processor core to the FPGA to control the hexapod. The project Xillybus IPs Core was designed to use four FIFOs, two focused on reading and two others on writing data. Each FIFO was configured to a 32 bit data width, a data transmission latency of 5 ms, a bandwidth of 10 MB/s, and a buffering time to autoset. The FPGA is internally forced to control the buffer RAM distribution for continuous reading and writing operations by configuring the buffering time to *autoset* and specifying the planned period for the maximal processor deprivation. The following equation gives the *RAM* size required for the DMA buffers' flow:

$$RAM = t \times BW \quad (1)$$

where *t* is the buffering time and BW is the expected data bandwidth.

For reading, all FIFOs must be empty and the enable signals (rd-en) activated. Thus, EEG data can fill the FIFOs until they all are full and the empty signal is disabled. Since FIFOs work with 32 bit and considering that the Emotiv EPOC+ device has a 14 bit resolution, a zero-padding operation was applied to each signal at the Most Significant Bit (MSB) position. Like the previous procedure, writing is enabled (wr-en at the high level) when all write FIFOs are empty (low level). Therefore, a finite state machine was designed to control the FIFOs' filling and emptying processes.

The EEG signal reading, processing, and classification algorithms were written in Python, the Verilog Language, the ANSI-C language (Nios® II Embedded Design Suite), and the Open Computing Language (OpenCL Standard) [28], which were tested and evaluated on the SoCKit. Table 1 summarizes the SoCKit resources used in the implemented experiments.

Table 1. SoCKit resources and materials deployed for the project implementation. HPS, Hard Processor System.

Label	Characteristics
SoCKit board	Altera Cyclone V SoC ARM, 5CSXFC6D6F31C8NES model, dual-core, ARM Cortex-A9 (HPS), 6 fractional Phase-Locked Loops (PLLs), 3.125 G transceivers
FPGA memories	1 GB (2 × 256 MB ×16) DDR3 SDRAM
HPS memories	1 GB (2 × 256 MB ×16) DDR3 SDRAM, 64 MB Quad Serial Peripheral Interface (QSPI) Flash
Display	24-bit VGA DAC; 128 × 64 dots LCD module
SD card image	64 GB, speed: Class 4
FPGA FIFO	Word length of 32 bit, transfer rate of 10 MB/s, filling time of 0.2 ms
Number of FIFOs	Four: two for downstream and two for upstream
DMA buffer memories	Autoset internals (automatic memory allocation)
Emotiv to FPGA streams	Delay time of 10 ms
Xillybus IPs Core latency	5 ms
FPGA to hexapod streams	Delay time of 10 ms
Clock frequencies	25, 50, and 100 MHz
FPGA configuration	Quad serial configuration device EPCQ256
Power consumption	1.023 W (internal power evaluation)
SSC-32 V2.0 card (Figure 4)	Channel servo controller, from 0.50 to 2.50 ms
USB-TTL adapter	USB to UART converter module

FPGA outputs were wire-connected to the hexapod servo-control board. The Central Pattern Generator (CPG), based on discrete-time neural networks, was adapted to move the hexapod robot [29].

The locomotion law defined by the CPGs and derived from the discrete-time spiking neuronal model [30] is mathematically described by:

$$V_i[k] = \gamma(1 - Z_i[k-1])V_i[k-1] + \sum_{j=1}^{12} W_{ij}Z_j[k-1] + I_i^{ext}, \quad (2)$$

where Z_i is the firing state of the ith neuron at time k, V_i is the potential membrane, W_{ij} is the synaptic influences (weights), I_i^{ext} is the external current, and γ is a dimensionless parameter. Mainly, $Z_i[k]$ is defined as a thresholded Heaviside function.

Moreover, considering that twelve servomotors control the hexapod movements, twelve neurons were required in this model; the input current was not needed (i.e., $I_i^{ext} = 0$), and $\gamma = 1$ to emulate a linear integrator.

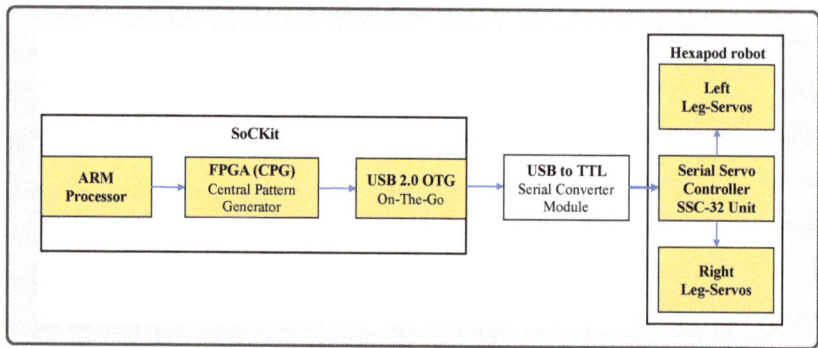

Figure 4. SoCKit-hexapod robot interconnection using a USB-TTL adapter and an SSC-32 V2.0 board.

2.3. BCI Dataset

The test subjects provided written consent to capture the EEG signals after carefully reading the experimental protocol to protect confidentiality. The specialized equipment used in the experiment was entirely commercial, not presenting any potential risk to the participants. Seven subjects were initially selected for the training process, and after addressing the defined paradigm, they followed an individual schedule.

Before and during each training session, the Emotiv Software Development Kit (Emotiv Xavier) monitored the subject's cognitive and emotional performances [15]. Hence, a dataset was created selecting four test subjects between 23 and 36 years old, trained and supervised to collect signals during several experimental tasks lasting three seconds each. According to the given task, subjects were instructed to stay still during the capture and invited to imagine closing and opening the right or the left fist focused on a stimulus video (Figure 5).

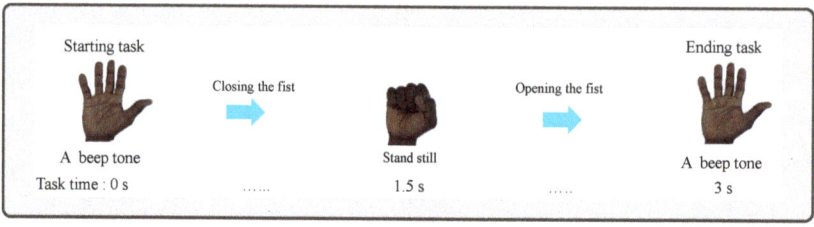

Figure 5. The MI task video serves as a subject stimulus. The task starts with a beep tone by closing the fist completely and opening the same fist until the end beep appears. Each task lasts 3 s.

The stimuli video of the fist closing-opening movements was played on the screen according to the temporal task sequence shown in Figure 6.

In the capture sequence, the first five seconds served to prepare the test subject, ending this phase with the audible Beep 1, followed by Task 1, related to the left fist MI action. The Beep 2 tone concludes this period and marks a pause of 3 s. Beep 3 triggers the end of this static period and starts a second preparation phase of 2 s. Beep 4 starts Task 2, related to the right fist MI task, ending with Beep 5. Therefore, the developed dataset consists of 2400 trials performed by four subjects (600 trials from each subject), representing 2400 × 19 s (12.67 h) of data capture. For each session duration, only signals of 3 s corresponding to Task 1 (left fist MI), 3 s to Task 2 (right fist MI), and 3 s to neutral action were gathered to build the dataset.

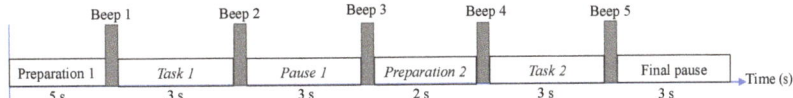

Figure 6. Time sequence of each trial. The total trial duration was established at 19 s, where 6 s were used for the related MI tasks: the left fist as Task 1 (3 s) and the right fist as Task 2 (3 s). The neutral or reference action was taken as the final pause (3 s) to have an equal number of samples per class.

2.4. Data Preprocessing

Signals from the F3, F4, FC5, and FC6 sensors were processed in the MI recognition process [31,32]. Such sensors were located in the rear portion of the frontal lobe, as shown in Figure 7.

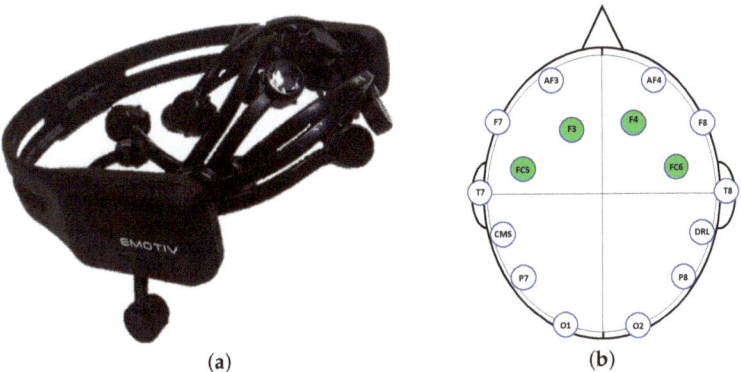

Figure 7. Electrodes' arrangement and nomenclature. The letter expresses the part of the brain where the sensor is placed, frontal (F), central (C), parietal (P), occipital (O), temporal (T), and frontoparietal (FP). Even numbers are used for the right hemisphere, while odd numbers for the left hemisphere. (**a**) Emotiv EPOC+ headset. (**b**) Electrode headset location.

The Emotiv EPOC+ headset was configured with three filters: a low-pass filter with a cutoff frequency at 85 Hz, an operational bandwidth between 0.16 and 43 Hz, and a band-rejection filter with a stop-band between 50 and 60 Hz [15,33]. According to the International EEG Waveform Society, the project paradigm is based on the *mu rhythm* processing, which occupies frequencies between 8 and 12 Hz [34]. The *mu rhythm* is the most used pattern in BCI systems considering the nature of the MI movements [35,36]. Thus, the mental imagery of body members' mobility can be perceived through the *mu rhythm* variations at the sensorimotor cortex, avoiding any real movement of the body limbs [37]. Lotze et al. determined that the left and right hands' physical movements cause an Event-Related Desynchronization (ERD) of the *mu rhythm* power, captured in different motor cortex areas [38]. Consequently, the F3 and FC5 sensors were selected for the left hemisphere, whereas F4 and FC6 for the right hemisphere on the sensorimotor cortex. Such a choice takes into account the sensor's closeness to the primary motor cortex location associated with the imagined and physical movements of the left and right hands [31,32].

2.5. MI-EEG Signals' Classification Based on a CNN-LSTM Architecture

Recurrent neural networks (e.g., LSTM networks) are composed of memory units that temporarily store information [39]. Such a network's layer structure is not unique because the interconnections between neurons are not based on a transportable (mutable) logic. The feature extraction and classification of EEG signals are done by combining two neural schemes, the CNN and LSTM. Figure 8 presents the CNN-LSTM architecture integrated into the SoCKit to decode robot commands. The overall network consists of a sequence of layers: a convolutional layer (CNN1), an LSTM layer (LSTM1), a convolutional layer (CNN2), followed by a max-pooling layer, a convolutional layer (CNN3), an LSTM layer (LSTM2), and a dense layer.

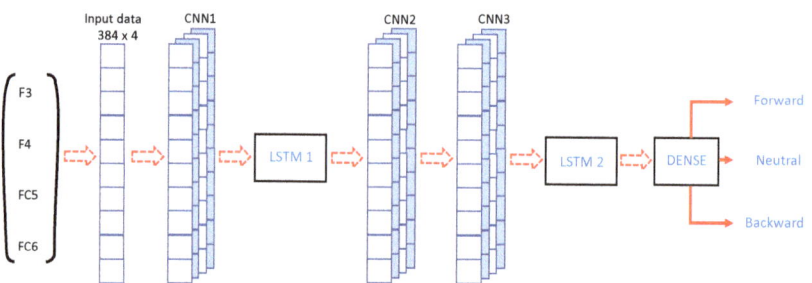

Figure 8. The proposed CNN-LSTM network architecture. A convolutional layer extracts features from preprocessed data, followed by an LSTM layer and two CNN layers, directly connected to a second LSTM layer and a dense layer to classify MI-EEG signals into the robot commands.

A 384 × 4 matrix was applied as the input to the CNN1 layer, which performed 32 convolutions with a 3 × 3 size kernels. In each convolutional layer (CNN1, CNN2, and CNN3), the padding parameter was configured to have the same temporal dimensions between input and output data. Weights were initialized according to a uniform distribution using the *He* initialization algorithm [40]. Dropout was applied to each convolutional layer with parameters tuned to 0.4, 0.2, 0.2, and 0.1 for the CNN1, CNN2, CNN3, and LSTM2 layers, respectively. According to the deep learning software interface Keras [41], for a dropout rate of 0.1, only 10% of the neurons are zeroed-out during the training phase, which reduces overfitting (overtraining).

On the other hand, LSTM layers contain 32 and 150 cells and receive feature matrices from convolutional layers for processing. The model was implemented in Keras and TensorFlow using the *categorical cross-entropy* loss function to evaluate the error between the estimated outputs and the ground-truth. The network was trained for 8000 epochs to meet the max accuracy, using the Nesterov-accelerated Adaptive Moment Estimation (NADAM) optimizer with a batch size of 512. A cyclical learning rate with a step-size of nine and minimum and maximum learning rates of 0.000001 and 0.0005, respectively, was used to speed up training [42]. The convolutional layers used the leaky Rectified Linear Unit (ReLU) as the activation function with $\alpha = 0.005$. This allowed obtaining a small non-zero gradient when a neuron has a negative net input. The leaky ReLU activation function $f(\alpha; x)$ is defined by:

$$f(\alpha; x) = \begin{cases} \alpha x & \text{if } x < 0 \\ x & \text{otherwise} \end{cases} \quad (3)$$

where α is a small positive constant [43]. However, *SoftMax* was used as the activation function of the fully connected layer following the LSTM2 layer to normalize the outputs, such that they may be interpreted as class probabilities [44].

Table 2 depicts the principal parameters of the proposed network model.

Table 2. Summary of the parameters of the CNN-LSTM architecture.

Layer	Filters/Cells/Rate	Output Shape	Parameters
CNN1	32	(None, 384, 32)	416
Dropout1	0.4	(None, 384, 32)	0
LSTM1	32	(None, 384, 32)	8320
CNN2	32	(None, 384, 32)	3104
Dropout2	0.2	(None, 384, 32)	0
Max-pooling	32	(None, 192, 32)	0
CNN3	32	(None, 192, 32)	3104
Dropout3	0.2	(None, 192, 32)	0
LSTM2	150	(None, 150)	109,800
Dropout 4	0.1	(None, 150)	0
Dense	1	(None, 3)	453

The convolutional layer CNN1 has only 416 parameters, while the CNN2 and CNN3 layers have 3104 parameters each. It must be highlighted that CNNs do not require a specially designed feature extraction stage because they can perform adaptive feature extraction directly on raw input data. Therefore, there were 125,197 parameters necessary for all layers. The output used a fully connected layer with *SoftMax* as the activation function, which produces the three class probabilities. During the neural network training, the neuron weights were randomly initialized using the *He* initialization algorithm.

3. Experimental Results

The proposed method's performance was evaluated according to the operative interconnection of the Emotiv EPOC+ headset, the SoCKit board, and the hexapod robot. It is worth noting that recognition algorithms were integrated into an embedded BCI system. MI-EEG recognition was achieved by implementing a CNN-LSTM architecture and creating, training, and validating an EEG dataset.

3.1. Qualitative Evaluation

Figure 9 shows the servomotor location and associated nomenclature, as well as the diagram of the hexapod locomotion sequence.

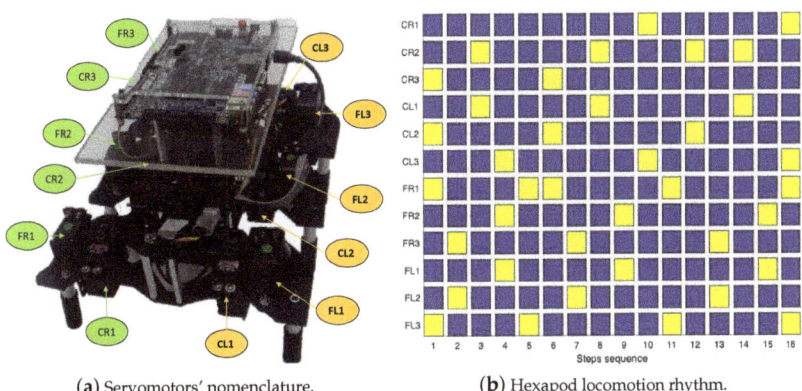

(**a**) Servomotors' nomenclature.   (**b**) Hexapod locomotion rhythm.

Figure 9. (**a**) The SoCKit-hexapod embedded system. The hexapod operates with twelve degrees of freedom using the Coxa (C) and Femur (F) articulations on the Right (R) and Left (L) sides. (**b**) Hexapod locomotion patterns. At any locomotion step, the corresponding servomotor activation is painted in yellow and deactivated in blue.

Figure 9b illustrates the repeating pattern sequences used to control the servomotors, which are integrated to move the hexapod synchronously. MI-EEG signals for closing-opening the right and left fists were processed by the SoCKit and transferred to the hexapod servomotors as commands to move forward, backward, and stop.

The dataset was constituted by signals captured from four chosen subjects, which were trained to reproduce each task pattern until they became familiar with the experience. The Xavier interface of Emotiv evaluated the subjects' mental state metrics before each training and capture [45], using cognitive, expressive, affective, and inertial sensors.

3.2. Quantitative Evaluation

A total of 2400 sessions were validated with four test subjects several times. This process allowed obtaining representative samples for the training and validation datasets. Therefore, two-thousand one-hundred sixty captures were used as the training data and 240 captures as the validation data. The number of classes was three (i.e., forward, backward, and neutral). Moreover, the stratified k-fold Cross-Validation (CV) was used with $k = 10$ to evaluate the system performance. Table 3 summarizes the dataset structure split into training and validation patterns.

Table 3. Data partition for training and validation.

	Dataset Structure	
Stratified k-fold CV	Training (90%)	Validation (10%)
$k = 10$	2160 captures	240 captures

The convolutional layer CNN1 used 416 parameters, while the CNN2 and CNN3 layers had 3104 parameters each. Thus, there were 125,197 parameters necessary for all layers.

3.2.1. EEG Signals' Dataset

The highlighted parameters in the created dataset were age, gender, and the capture sequence duration. The dataset features are summarized in Table 4.

Table 4. Local dataset description.

Order	ID	Age	Gender
1	AB	34	Male
2	CD	23	Female
3	EF	36	Male
4	GH	23	Female

The subjects of the dataset required significant mental focus on the stimulus to get reliable signals, which was solved with extensive experimentation for easy adaptation to this experience.

3.2.2. Model Evaluation

The proposed framework's evaluation considers the accuracy of the hexapod movement commands according to the three predefined tasks identified from the MI signals. Simultaneously, the designed BCI must optimize real-time signal processing. The neural network training was repeated five times, achieving an average accuracy of 84.69% for the three experimental predefined tasks. The highest accuracy of 87.6% was obtained for the sixth k-fold iteration, whereas the lowest accuracy of 81.88% was found with the second k-fold iteration. The created database provides signal patterns lasting six hours collected from four test subjects.

Table 5 shows the classification accuracy for each test subject, while Table 6 summarizes the accuracy reached with different subject combinations.

Table 5. Classification accuracy for each dataset subject.

Subjects	Training Accuracy	Test Accuracy
AB	95.49%	82.38%
CD	94.92%	80.83%
EF	99.03%	83.54%
GH	97.38%	85.25%

Table 6. Classification accuracy of subject combinations.

Subjects	Training Accuracy	Test Accuracy
AB-CD	98.9%	83.7%
AB-CD	99.2%	81.3%
AB-EF	99.8%	82.7%
AB-GH	96.5%	79.5%
CD-EF	99.4%	85.7%
CD-GH	91.1%	83.6%
EF-GH	99.5%	84.2%
AB-CD-EF	99.7%	83.4%
AB-CD-GH	99.8%	85.4%
CD-EF-GH	99.6%	86.1%

This study was additionally evaluated with the Scientific-Data dataset [27]. The Scientific-Data gathers five EEG signal paradigms, captured with the Nihon Kohden Neurofax EEG-1200 electroencephalograph and the JE-921A amplifier. Scientific-Data signals were selected to evaluate our dataset. The proposed BCI evaluation is based on left- and right-hand motor imagining of closing and opening the respective fist once, defined as Paradigm #1 [27]. After imagining such actions, the participant remained passive until the next action signal was presented.

Table 7 shows that the average accuracy achieved for the three task classification was 79.2% using the evaluated Scientific-Data set.

Table 7. Accuracies achieved with our and Scientific-Data datasets.

Reference	Dataset Type	Brain Signals	Accuracy
Proposed method	Our dataset (4 subjects)	MI-EEG	84.69%
Scientific-Data [27]	Public dataset (7 subjects)	MI-EEG	79.2%

It is worth mentioning that the proposed network model was trained on a computer with an NVIDIA GeForce® GTX 1080 GPU. Next, the descriptive files (weight files, module files) were migrated to the SoCKit card to optimize the processing latency time of the EEG signals. The embedded system took approximately 0.750 s to decide whether the samples present referred to the right-fist, left-fist, or none using the FIFO configuration and the Emotiv EPOC+ rate of 128 Samples Per Second (SPS). The processing time with the Scientific-Data dataset (sampled at 200 SPS) was evaluated as 0.279 s with signals from the C3, C4, and Cz sensors.

Moreover, two closely related embedded BCI approaches [19,20] were also included to compare the proposed method in Table 8, including the signals' processing time and the number of channels.

Table 8. Processing time comparison with two additional embedded-BCI systems.

Method	Embedded-BCI System	Number of Channels	Running Time
Lin et al. [20]	Embedded Brain computer interface-based Smart Living Environmental Auto-adjustment Control System (BSLEACS)	1	2000 ms
Belwafi et al. [19]	Hardware/software prototype of EEG-based BCI system	22	399 ms
Proposed method	Embedded SoCKit	4 (Emotiv)	750 ms
Proposed method	Embedded SoCKit	3 (Scientific-Data)	279 ms

Therefore, this work's contributions are developing an MI-EEG dataset, a CNN-LSTM model implemented in the FPGA SoC board for real-time signal processing, and the embedded BCI architecture implementation with different technologies.

3.3. Discussion

The proposal was implemented on the Altera FPGA SoC environment, building the respective hardware design modules shown in Figure 1. A mechanism to read EEG signals in real time on the SoCKit board was designed with a delay limit of about 10 ms for the Emotiv EPOC+ capture. A proper buffer module synchronization guaranteed the regularity of data being sent to the processor, where the available memory (DMA) was dynamically allocated.

Quartus II Version 13.0 was used to embed the project descriptive architectures, to ensure the processing and classification of motor imagining EEG signals. Real-time signal processing starts when the Emotiv EPOC+ system sends EEG signals at 128 samples per second to the SoCKit card. Next, the NIos-II IDE cache memory was previously preloaded with the designed neural network parameters. Therefore, the SoCKit can instantly convert classified EEG signal features into commands to move the hexapod. Moreover, the SoCKit USB 2.0 On-The-Go (OTG) port was used to connect the Serial Servo Controller SSC-32 Control board through the USB to TTL serial converter module.

Note that physically, the SoCKit board was embedded in the hexapod, and they moved together.

Table 5 shows the achieved variability of the accuracy results with each subject, confirming intrinsic differences among EEG signal characteristics [46], besides the closely related faithful reproduction by each subject. In Table 6, different combinations of the subjects' signals are given to appreciate the data classification accuracy according to the dataset size variation. Thus, a high test accuracy of 85.7% was reached by combining two signal groups from the CD and EF subjects while combining three signal groups from the CD, EF, and GH subjects; the highest score achieved was 86.1%.

4. Conclusions

This study presents the development of an embedded BCI system based on EEG signals, corresponding to imagined fist movements, captured with an Emotiv EPOC+ headset and processed in real time on the SoCKit Cyclone V SoC card to control a hexapod robot. The designed framework allows controlling the forward and backward movements of a hexapod robot, using two MI tasks: closing and opening the right fist, closing and opening the left fist, and the neutral (reference) action.

Likewise, an MI-EEG dataset is created. Other hexapod locomotion modalities, including turning right or left, running, sitting, or climbing, are planned for future work, considering this framework as the first experimental and incremental approach. MI-EEG signal recognition was carried out using a hybrid CNN-LSTM. By using stratified 10-fold cross-validation, the average task accuracy is determined as 84.69%.

The digital logical design guarantees adequate functionality in the integral transmission of data. Hence, FIFOs communicating with FPGA outputs are implemented at a 32 bit word-length, running at 10 MB/s. Task recognition delay on the SoCKit is estimated at 755 ms and about 500 ms in executing the hexapod movements, including intrinsic delays in the SSC-32 V2.0 card. This study proves the active and accurate locomotion of a hexapod robot, exploiting EEG brain signals captured by an Emotiv EPOC+ headset and processed by a SoCKit card using a pre-trained CNN-LSTM as a classifier. The research perspectives of this study primarily include building a larger and robust database with more sophisticated sensing equipment, increasing the number of subjects, and modular tasks to control the mobile robot. Likewise, the proposed methodology could be straightforwardly applied using EEG signals to control special devices. In such a context, the EEG based BCI of a wheelchair will be analyzed to support human mobility.

Author Contributions: Conceptualization, T.M.-V.; data curation, H.R.-G. and J.M.C.-D.; formal analysis, H.R.-G. and M.A.I.-M.; funding acquisition, M.A.I.-M.; investigation, T.M.-V., J.R.-P., J.M.C.-D., and J.G.A.-C.; methodology, J.R.-P. and J.G.A.-C.; software, T.M.-V., H.R.-G., and J.G.A.-C.; validation, J.R.-P. and M.A.I.-M.; writing—original draft, T.M.-V.; writing—review and editing, J.R.-P., J.M.C.-D., and J.G.A.-C. All authors read and agreed to the published version of the manuscript.

Funding: This project was supported by the Mexican National Council of Science and Technology CONACyT, under Grant Number 763527/600853 and the University of Guanajuato.

Institutional Review Board Statement: The study was conducted according to the guidelines of the Declaration of Helsinki for procedures involving human participants. Ethical review and approval are waived for this kind of study. In EEG signals' capture, no invasive procedures were involved, nor were special biodata captured that could be used to identify the participants.

Informed Consent Statement: Formal written consent to capture EEG signals from participants is available on demand.

Data Availability Statement: The dataset presented in this study is publicly available in Mendeley Data repository at doi: 10.17632/2sdzh8dgvt.1, https://data.mendeley.com/datasets/2sdzh8dgvt/draft?preview=1, accessed on 7 February 2021.

Acknowledgments: This project was fully supported by the Electrical and Electronics Departments of the Universidad de Guanajuato under the Program POA 2021, Grant 145790, The Mexican National Council of Science and Technology (CONACYT) through the grant 763527/600853, and the research project ERC-297702.

Conflicts of Interest: The authors declare no conflict of interest.

References

1. Jorge, J.; Van der Zwaag, W.; Figueiredo, P. EEG-fMRI integration for the study of human brain function. *NeuroImage* **2014**, *102*, 24–34. [CrossRef]
2. Jacobs, J.; Miller, J.; Lee, S.; Coffey, T.; Watrous, A.; Sperling, M.; Sharan, A.; Worrell, G.; Berry, B.; Lega, B.; et al. Direct Electrical Stimulation of the Human Entorhinal Region and Hippocampus Impairs Memory. *Neuron* **2016**, *92*, 983–990. [CrossRef]
3. Ieracitano, C.; Mammone, N.; Hussain, A.; Morabito, F. A novel multi-modal machine learning based approach for automatic classification of EEG recordings in dementia. *Neural Netw.* **2020**, *123*, 176–190. [CrossRef]
4. Safi, M.; Safi, S. Early detection of Alzheimer's disease from EEG signals using Hjorth parameters. *Biomed. Signal Process. Control* **2021**, *65*, 102338. [CrossRef]
5. Ahmedt-Aristizabal, D.; Fernando, T.; Denman, S.; Robinson, J.; Sridharan, S.; Johnston, P.; Laurens, K.; Fookes, C. Identification of Children at Risk of Schizophrenia via Deep Learning and EEG Responses. *IEEE J. Biomed. Health Inform.* **2021**, *25*, 69–76. [CrossRef] [PubMed]
6. Tong, S.; Thakor, N.V. *Quantitative EEG Analysis Methods and Clinical Applications*; Artech House: Boston, MA, USA, 2009.
7. Markopoulos, K.; Mavrokefalidis, C.; Berberidis, K.; Daskalopoulou, E. BCI based approaches for real-time applications. In Proceedings of the ACM International Conference Proceeding Series, Patras, Greece, 10 November 2016; pp. 1–6. [CrossRef]
8. Azmy, H.; Safri, N. EEG based BCI using visual imagery task for robot control. *J. Teknol. (Sci. Eng.)* **2013**, *61*, 7–11. [CrossRef]
9. Palankar, M.; Laurentis, K.D.; Dubey, R. Using biological approaches for the control of a 9-DoF wheelchair-mounted robotic arm system: Initial experiments. In Proceedings of the IEEE International Conference on Robotics and Biomimetics, Bangkok, Thailand, 22–25 February 2009; pp. 1704–1709. [CrossRef]

10. Palmer, L.; Palankar, M. Blind hexapod walking over uneven terrain using only local feedback. In Proceedings of the IEEE International Conference on Robotics and Biomimetics, Karon Beach, Thailand, 7–11 December 2011; pp. 1603–1608. [CrossRef]
11. Karimi, D.; Nategh, M. Kinematic non-linearity analysis in hexapod machine tools: Symmetry and regional accuracy of workspace. *Mech. Mach. Theor* **2014**, *71*, 115–125. [CrossRef]
12. Shashibala, T.; Gawali, B.W. Implementation of Robotic arm control with Emotiv EPOC. *Int. J. Adv. Eng. Res. Sci.* **2016**, *3*, 22–26.
13. Dewangga, S.; Tjandrasa, H.; Herumurti, D. Robot motion control using the Emotiv EPOC EEG system. *Bull. Electr. Eng. Inform.* **2018**, *7*, 279–285. [CrossRef]
14. Dewald, K.; Jacoby, D. Signal Processing In Embedded Systems. *Lat. Am. Trans. IEEE (Rev. IEEE Am. Lat.)* **2013**, *11*, 664–667. [CrossRef]
15. Salgado, J.; Barrera, C.; Monje, B. Emotiv EPOC BCI with Python on a Raspberry Pi. *Sist. Telemát.* **2016**, *14*, 27–38. [CrossRef]
16. Tejwani, K.; Vadodariya, J.; Panchal, D. Biomedical Signal Detection using Raspberry Pi and Emotiv EPOC. In Proceedings of the 3rd International Conference on Multidisciplinary Research & Practice (IJRSI), Ahmedabad Gujarat, India, 24 December 2016; Volume V, pp. 178–180.
17. Chen, J.; Sun, H. The Digital Signal Processing Algorithm Implemented on ARM Embedded System. *Adv. Mater. Res.* **2013**, *756–759*, 3958–3961. [CrossRef]
18. Haitham, A.; Khalaf, O.; Bazel, B.; Haithem, M. Wearable Ambulatory Technique for Biomedical Signals Processing Based on Embedded Systems and IoT. *Int. J. Adv. Sci. Technol.* **2020**, *29*, 360–371.
19. Belwafi, K.; Ghaffari, F.; Djemal, R.; Romain, O. A Hardware/Software Prototype of EEG based BCI System for Home Device Control. *J. Signal Process. Syst.* **2017**, *89*, 263–279. [CrossRef]
20. Lin, C.T.; Lin, B.S.; Lin, F.C.; Chang, C.J. Brain Computer Interface-Based Smart Living Environmental Auto-Adjustment Control System in UPnP Home Networking. *Syst. J. IEEE* **2014**, *8*, 363–370. [CrossRef]
21. Palumbo, A.; Amato, F.; Calabrese, B.; Cannataro, M.; Cocorullo, G.; Gambardella, A.; Guzzi, P.H.; Lanuzza, M.; Sturniolo, M.; Veltri, P.; et al. An Embedded System for EEG Acquisition and Processing for Brain Computer Interface Applications. In *Wearable and Autonomous Biomedical Devices and Systems for Smart Environment: Issues and Characterization*; Springer: Berlin/Heidelberg, Germany, 2010; pp. 137–154. [CrossRef]
22. Lin, C.T.; Chen, Y.C.; Huang, T.Y.; Chiu, T.T.; Ko, L.W.; Liang, S.F.; Hsieh, H.Y.; Hsu, S.; Duann, J.R. Development of Wireless Brain Computer Interface With Embedded Multitask Scheduling and Its Application on Real-Time Drivers' Drowsiness Detection and Warning. *IEEE Trans. Bio-Med. Eng.* **2008**, *55*, 1582–1591. [CrossRef]
23. Lin, J.S.; Huang, S. An FPGA-Based Brain-Computer Interface for Wireless Electric Wheelchairs. *Appl. Mech. Mater.* **2013**, *284–287*, 1616–1621. [CrossRef]
24. Yang, J.; Yao, S.; Wang, J. Deep Fusion Feature Learning Network for MI-EEG Classification. *IEEE Access* **2018**, *6*, 79050–79059. [CrossRef]
25. Bresch, E.; Großekathöfer, U.; Garcia-Molina, G. Recurrent Deep Neural Networks for Real-Time Sleep Stage Classification From Single Channel EEG. *Front. Comput. Neurosci.* **2018**, *12*, 85. [CrossRef]
26. Torres, L.; Imamoğlu, N.; Gonzalez-Torres, A.; Kouyama, T.; Kanemura, A. Evaluation of neural networks with data quantization in low power consumption devices. In Proceedings of the 2020 IEEE 11th Latin American Symposium on Circuits Systems (LASCAS), San Jose, Costa Rica, 25–28 February 2020; pp. 1–4. [CrossRef]
27. Kaya, M.; Binli, M.; Ozbay, E.; Yanar, H.; Mishchenko, Y. A large electroencephalographic motor imagery dataset for electroencephalographic brain computer interfaces. *Sci. Data* **2018**, *5*, 180211. [CrossRef]
28. Intel. Intel FPGA SDK for OpenCL Standard Edition. 2018. Available online: https://www.intel.com/content/dam/www/programmable/us/en/pdfs/literature/hb/opencl-sdk/aocl_c5soc_getting_started.pdf (accessed on 7 February 2021).
29. Guerra-Hernandez, E.; Espinal, A.; Batres-Mendoza, P.; Garcia-Capulin, C.; Romero-Troncoso, R.; Rostro-Gonzalez, H. A FPGA-Based Neuromorphic Locomotion System for Multi-Legged Robots. *IEEE Access* **2017**, *5*, 8301–8312. [CrossRef]
30. Soula, H.; Beslon, G.; Mazet, O. Spontaneous dynamics of asymmetric random recurrent spiking neural networks. *Neural Comput.* **2006**, *18*, 60–79. [CrossRef] [PubMed]
31. Kato, K.; Takahashi, K.; Mizuguchi, N.; Ushiba, J. Online detection of amplitude modulation of motor-related EEG desynchronization using a lock-in amplifier: Comparison with a fast Fourier transform, a continuous wavelet transform, and an autoregressive algorithm. *J. Neurosci. Methods* **2018**, *293*, 289–298. [CrossRef]
32. Alomari, M.; Awada, E.; Younis, O. Subject-Independent EEG-Based Discrimination Between Imagined and Executed, Right and Left Fists Movements. *Eur. J. Sci. Res.* **2014**, *118*, 364–373.
33. Badcock, N.A.; Mousikou, P.; Mahajan, Y.; de Lissa, P.; Thie, J.; McArthur, G. Validation of the Emotiv EPOC® EEG gaming system for measuring research quality auditory ERPs. *PeerJ* **2013**, *1*, e38. [CrossRef]
34. Xu, B.; Fu, Y.; Shi, G.; Yin, X.; Wang, Z.; Li, H.; Jiang, C. Enhanced Performance by Time-Frequency-Phase Feature for EEG based BCI Systems. *Sci. World J.* **2014**, *2014*, 420561. [CrossRef]
35. Pfurtscheller, G.; Brunner, C.; Schlagl, A.; da Silva, F.L. Mu rhythm (de)synchronization and EEG single-trial classification of different motor imagery tasks. *NeuroImage* **2006**, *31*, 153–159. [CrossRef]
36. Fu, R.; Tian, Y.; Bao, T.; Meng, Z.; Shi, P. Improvement Motor Imagery EEG Classification Based on Regularized Linear Discriminant Analysis. *J. Med. Syst.* **2019**, *43*, 1–13. [CrossRef] [PubMed]

37. Hwang, H.J.; Kwon, K.; Im, C.H. Neurofeedback based motor imagery training for brain-Computer interface (BCI). *J. Neurosci. Methods* **2009**, *179*, 150–156. [CrossRef]
38. Lotze, M.; Montoya, P.; Erb, M.; Hülsmann, E.; Flor, H.; Klose, U.; Birbaumer, N.; Grodd, W. Activation of cortical and cerebellar motor areas during executed and imagined hand movements: An fMRI study. *J. Cogn. Neurosci.* **1999**, *11*, 491–501. [CrossRef]
39. Wang, P.; Jiang, A.; Liu, X.; Shang, J.; Zhang, L. LSTM based EEG Classification in Motor Imagery Tasks. *IEEE Trans. Neural Syst. Rehabil. Eng.* **2018**, *26*, 2086–2095. [CrossRef] [PubMed]
40. Yam, J.; Chow, T. A weight initialization method for improving training speed in feedforward neural network. *Neurocomputing* **2000**, *30*, 219–232. [CrossRef]
41. Chollet, F. Keras. GitHub. 2015. Available online: https://github.com/fchollet/keras (accessed on 7 February 2021).
42. Smith, L.N. Cyclical Learning Rates for Training Neural Networks. In Proceedings of the 2017 IEEE Winter Conference on Applications of Computer Vision (WACV), Santa Rosa, CA, USA, 24–31 March 2017; pp. 464–472. [CrossRef]
43. Xu, B.; Wang, N.; Chen, T.; Li, M. Empirical Evaluation of Rectified Activations in Convolutional Network. *arXiv* **2015**, arXiv:1505.00853.
44. Gao, B.; Pavel, L. On the Properties of the Softmax Function with Application in Game Theory and Reinforcement Learning. *arXiv* **2017**, arXiv:1704.00805.
45. Paszkiel, S.; Szpulak, P. Methods of Acquisition, Archiving and Biomedical Data Analysis of Brain Functioning. *Adv. Intell. Syst. Comput.* **2018**, *720*, 158–171. [CrossRef]
46. Hu, L.; Zhang, Z. *EEG Signal Processing and Feature Extraction*; Springer: Singapore, 2019; [CrossRef]

Article

Artificial Neural Network, Quantile and Semi-Log Regression Modelling of Mass Appraisal in Housing

Jose Torres-Pruñonosa [1,*], Pablo García-Estévez [2] and Camilo Prado-Román [3]

1. Facultad de Empresa y Comunicación, Universidad Internacional de la Rioja, 26006 Logroño, Spain
2. Colegio Universitario de Estudios Financieros (CUNEF), 28040 Madrid, Spain; pgestevez@cunef.edu
3. Department of Business Economics, Universidad Rey Juan Carlos, 28933 Madrid, Spain; camilo.prado.roman@urjc.es
* Correspondence: jose.torresprunonosa@unir.net

Abstract: We used a large sample of 188,652 properties, which represented 4.88% of the total housing stock in Catalonia from 1994 to 2013, to make a comparison between different real estate valuation methods based on artificial neural networks (ANNs), quantile regressions (QRs) and semi-log regressions (SLRs). A literature gap in regard to the comparison between ANN and QR modelling of hedonic prices in housing was identified, with this article being the first paper to include this comparison. Therefore, this study aimed to answer (1) whether QR valuation modelling of hedonic prices in the housing market is an alternative to ANNs, (2) whether it is confirmed that ANNs produce better results than SLRs when assessing housing in Catalonia, and (3) which of the three mass appraisal models should be used by Spanish banks to assess real estate. The results suggested that the ANNs and SLRs obtained similar and better performances than the QRs and that the SLRs performed better when the datasets were smaller. Therefore, (1) QRs were not found to be an alternative to ANNs, (2) it could not be confirmed whether ANNs performed better than SLRs when assessing properties in Catalonia and (3) whereas small and medium banks should use SLRs, large banks should use either SLRs or ANNs in real estate mass appraisal.

Keywords: artificial neural networks; banking; hedonic prices; housing; quantile regression

1. Introduction

The excessive dependence on the real estate industry, in addition to the softening of credit standards [1], meant that the economic and financial crisis of the end of the first decade of the 21st century hit Spain more severely than other developed economies. Consequently, 61,495 million euros were needed to bail out the banking system, which has been radically transformed by means of mergers, acquisitions and the transformation of almost all savings banks into commercial banks [2,3]. Spanish financial institutions have suffered during this crisis, as there has been a significant rise in high-risk mortgages and properties being valued at their historical value. Hence, one of the biggest challenges the banking sector has faced in recent years has been finding the best way to value this stock. An optimal valuation has two advantages: first, it helps to know the real financial situation of the bank; second, if the property is assessed according to the market, it can be sold in a shorter period.

The hedonic analysis is an approach that is widely used to deal with the heterogeneity involved in valuing housing. The hedonic price methodology is used to explain the price of heterogeneous products with heterogeneous characteristics by noting that the implicit marginal price of these characteristics can be found out by means of estimating models that explain the price based on the product's characteristics. The economic literature that deals with hedonic prices arose in the context of the car market. This was the framework for the classical work by Griliches [4], who made these models popular by estimating car prices after controlling the characteristic that affected their prices, such as fuel consumption and horsepower. Nonetheless, there was a previous paper in the early 1940s that can be considered the first one to deal with the hedonic price methodology [5]. Once the technique became popular in the 1950s [6], more than a decade was necessary to establish its theoretical framework. In this regard, Rosen [7] provided the theoretical foundation by means of showing how marginal prices are implicitly determined by the characteristics of heterogeneous products that can be estimated by means of a model (called the hedonic price model), which explains the price of products based on their characteristics (the hedonic technique is based on modern consumer choice theory; this theory states that a consumer does not obtain utility directly from the good but from its characteristics [8]). Certainly, real estate is a type of good that fits perfectly into the hedonic price models framework since each house is a unique good because each dwelling is somehow different from the rest of them. There are many examples of hedonic studies of the housing market [9–22]. Hedonic prices can also be estimated using quantile regressions (QRs). When the estimation of the conditional mean cannot capture the links between the explanatory variables and the dependent variable throughout the whole distribution of the latter, QRs are frequently used. QRs have also recently been used in the literature on housing economics [23–36].

A problem arises because the hedonic price function is generically nonlinear. Therefore, the quantity of the characteristic, as well as its marginal implicit price, are endogenous in the hedonic price model (selecting a suitable nonlinear specification for the hedonic price function also solves this problem; in this regard, Ekeland et al. [37] stated that the demand parameters are always detected in single-market data if the marginal price function is nonlinear, which is called a "generic property of equilibrium in the hedonic model"). Due to their functional flexibility, artificial neural networks (ANNs) have been proposed as a means of extracting the nonlinear structures underlying the hedonic pricing approach. Given that parametric estimation in an ANN does not depend on the range of a regressor matrix, ANNs are better than the models that need to use large sets of dummies, and Selim [38] supports that ANNs are better estimators than traditional models.

Since the first work in this area by White [39], there is an abundance of literature about the application of neural networks to real estate prices. Frequently, these studies compare the results of an ANN to traditional (parametric) regression models. In some papers, ANNs perform better [40–46]. In other papers, standard hedonic regressions perform as well as the best ANN [47–49]. Other papers condition the utility of neural networks to the accomplishment of certain variables. Nghiep and Cripps [50] determined that ANNs obtain better results than regression models when a large sample is used; Liu et al. [51] demonstrated that fuzzy neural models have the ability to approximate and are useful to estimate prices, but this is dependent on the database quality. Do and Grudnitski [41] examined the effect of age on housing by means of neural networks and found a negative relationship between value and age, but only during the first 16 to 20 years; then prices increase. McGreal et al. [48] used neural networks and asserted that better results are obtained when postal code is used as a delimiter. Peterson and Flanagan [45] affirmed that ANNs generate a smaller valuation error than other models and that their out-of-sample pricing precision is greater. Recent papers analysed hedonic variables using ANNs (e.g., [52–54]).

A literature gap in regard to the comparison between ANN and QR modelling of hedonic prices in housing has been identified. A search in the Web of Science Core Collection was carried out in order to confirm this. In fact, the following Boolean search of terms in the title, abstract or keywords (TS) was used: TS = (housing) AND TS = ("neural network*") AND TS = ("quantile regression*"). In other words, a combination of "housing" along with "neural network*" and "quantile regression*" was searched. Only two papers were found [55,56]. Neither of these papers compared ANN and QR modelling when assessing housing prices. As such, our article is the first paper to include this comparison. Furthermore, taking into account that many papers [57–59] have compared the performance of semi-log regression (SLR) modelling of hedonic prices in the housing market with other models, SLR modelling was included in our study as a benchmark to compare the results obtained. Therefore, a comparison between ANN, SLR and QR modelling performances in terms of the goodness of fit and estimation ability was carried out.

A second contribution to the field has to do with the size of the sample and the market analysed. Previous papers have addressed and demonstrated that there is a better performance of ANNs in comparison with hedonic prices. Even some of these analyses have been carried out in Spain. Nonetheless, these studies did not have a big sample and usually tended to analyse only a town [60] or even a single district of a city. Take the example of Tabales et al. [61] with a sample of 2888 dwellings in the city of Córdoba. Likewise, Tabales et al. [62] analysed 102 commercial premises in the same city. In a similar way, Baldominos et al. [63] performed the analysis with 2266 real estates in Salamanca district in Madrid city. Landajo et al. [43] performed their analysis in a Spanish region, Asturias, but with a sample of only 364 apartments. This study contributes to the research field because the sample used consisted of 188,652 dwellings split into two sub-samples, with the smallest one of these sub-samples (n = 24,781) being higher than all the samples used in the papers previously mentioned in this paragraph. Furthermore, this is the first study that dealt with the Catalan housing market as a whole, given that the sample used represents 4.88% of the total number of housing stock in Catalonia (Nomenclature des Unités Territoriales Statistiques II (NUTS-II)) from 1994 to 2013.

Third, we aimed to identify which real estate mass valuation method, out of the three models analysed, is more suitable to be used by banks. On the one hand, it is true that there is a bias between appraisals and transaction prices [64]. Nonetheless, when it comes to mortgages, appraisals are the only prices available for banks in Spain, including the Spanish Central Bank (Banco de España) [1,65]. In this regard, the Spanish Central Bank obliges banks to carry out periodical real estate mass appraisals [45,66] of all their properties that have been used as mortgage collaterals in order to quantify their potential impairment losses. In this context, appraisals are accepted to conduct real state mass valuation models. On the other hand, in Spain, the size of financial intermediaries ranges from very large (mainly commercial banks) to very small (mainly cooperatives and savings banks). This fact has intensified over the last few decades by means of different waves of mergers and acquisitions, some of which were motivated by the transformation of savings banks into commercial banks [67,68] and others were due to concentration processes in view of the increase in productivity. In fact, Spain seems to be currently immersed in a new process of acquisitions and mergers that will create a completely different banking ecosystem. Should all Spanish banks use the same real estate mass appraisal model regardless of their size? It is true that the datasets used in this article are limited to Catalonia. However, the Catalan real estate market represents the Spanish market well [69,70]. To begin with, according to the official statistics published by the Statistical National Institute of Spain (Instituto Nacional de Estadística, hereinafter referred to as INE) [71], Catalonia represents 15.33% of the Spanish housing stock and 14.39% of the real estate transactions. Furthermore, due to its demographic heterogeneity, Catalonia is a representative region of Spain as a whole. On the one hand, the second-largest city in Spain, Barcelona (NUTS-V), is the administrative capital of Catalan and ten cities located in Catalonia are among the 50 largest cities in Spain that are not a province (NUTS-III) capital (according to INE). On the other hand, Catalonia has many rural areas due to it being the Spanish autonomous community (NUTS-II) with the third-most trees per hectare according to the Minister for the Ecological Transition and Demographical Challenge (Ministerio para la Transición Ecológica y el Reto Demográfico) [72], having mountainous areas and having the fifth-most kilometres of coastline according to the Geographical National Institute (Instituto Geográfico Nacional) [73]. Finally, the performance of the Catalan and the Spanish housing market are homogenous. For instance, the price per square meter in the free market of dwellings (Figure 1) calculated since 1995 by the Ministry of Transport, Mobility and Urban Agenda (Ministerio de Transportes, Movilidad y Agenda Urbana) [74] shows that in both Catalonia and Spain, the increase of prices extended through to 2008, with a marked growth since the beginning of the century, prices dramatically fell through to 2014 and, thereafter, started a moderate increase. In fact, the correlation of these prices between Catalonia and Spain is significant (<0.01), with a Fisher correlation coefficient of 0.995. Likewise, similar trends are shown in Figure 2 in regard to the Housing Price Index (Índice del Precio de la Vivienda) published by INE [71] since 2007, with a significant ($p < 0.01$) correlation of 0.986. As such, Catalonia can be considered a representative housing market of Spain.

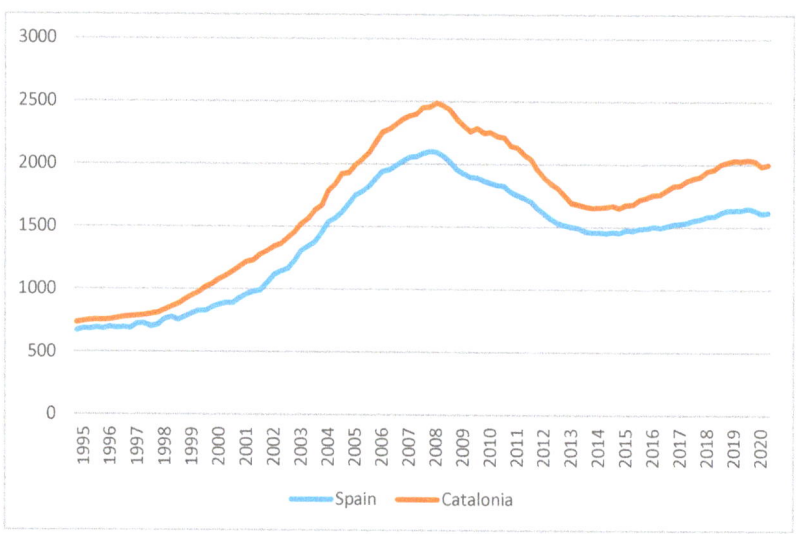

Figure 1. Price per square meter in the free market of dwellings (1995–2020). Source: Ministry of Transport, Mobility and Urban Agenda (Ministerio de Transportes, Movilidad y Agenda Urbana) [74].

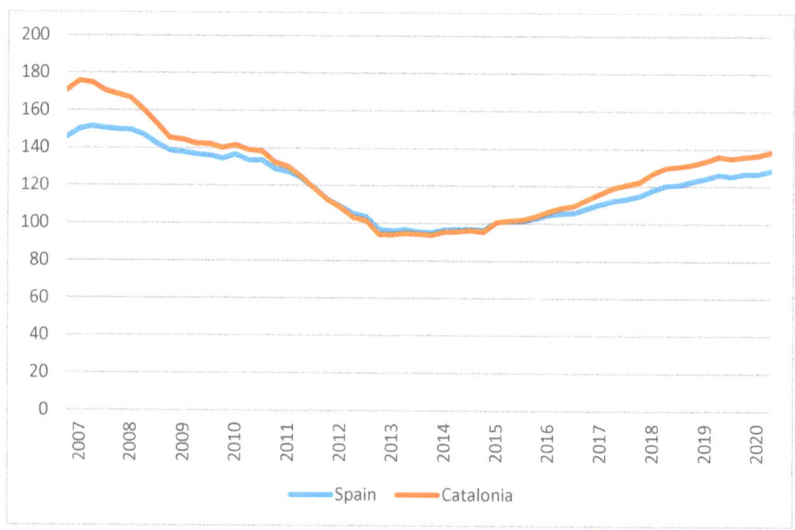

Figure 2. Housing Price Index (2007–2020). Source: Instituto Nacional de Estadística (INE) [71].

Overall, the aim of this study was threefold in that it focused on answering the following three questions: (1) Is QR valuation modelling of hedonic prices in the housing market an alternative to ANN? (2) Do housing assessments in the case of Catalonia confirm that ANNs produce better results than SLRs, as it does in other markets? (3) Out of the three analysed models, should all Spanish banks use the same mass appraisal model to assess real estate? In this paper, we present new evidence to compare the performance of QR and SLR hedonic models relative to ANN modelling using the data of properties owned by two banks.

The paper is structured as follows: the methodologies that were used are analysed in Section 2. The datasets used and the analysed variables are described in Section 3. Thereafter, the performance results of the models created are shown and discussed in Section 4. Finally, Section 5 includes the main conclusions and recommendations about the valuation methods to be used by banks and proposals for further research.

2. Methodology

Three methods were used in this study in order to value properties: ANNs, SLRs and QRs.

Neural networks are universal approximators of functions [75–77] and are used to adjust functions and also to estimate results. Even though the inception of ANNs can be found in the 1960s [78,79], they became more prevalent at the end of the last century as an alternative to the predominant Boolean logical computation [80].

Neural networks are based on an artificial neuron, which processes data in a similar way to a biological neuron named a perceptron [81]. Even though a single neuron cannot undertake a logical process on its own, it is possible for a group of them to do it. This is the reason why neurons are grouped in layers such that they can be used to make logical calculations in networks. A typical neural network has three layers. The first one works as data input. In the second one, which is hidden, data are processed. The third one works as data output. Every single neuron in a layer is connected to every neuron of the following layer via synaptic weights. Hence, when a neuron obtains a result, it is sent to all the neurons in the following layer [82] (see Figure 3).

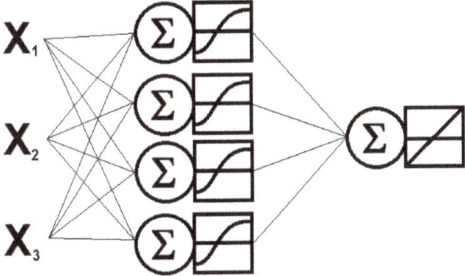

Figure 3. Representation of a neural network with three input data neurons, four hidden ones and one output data neuron

The most used supervised neural network is known as a multilayer perceptron (MLP) [75,83]. It consists of a three-layer network (input, hidden and output) that uses sigmoid functions as the transference function in the hidden layer.

The basis of this model is the artificial neuron (AN). It is a mathematical representation of a biological neuron. A representation of a perceptron is given in Figure 4. The AN receives inputs $(X_1, X_2, \ldots, X_N)$ and these inputs are weighted $(W_1, W_2, \ldots, W_3)$. When the sum-product of the inputs and weights exceeds a threshold (θ_i), the exceeded part is the input of the transfer function. This function is usually a sigmoid (Equation (1)) or tan-sigmoid function (Equation (2)).

$$f(x) = \frac{1}{1+e^{-x}} \qquad (1)$$

$$f(x) = \frac{2}{1+e^{-2x}} - 1 \qquad (2)$$

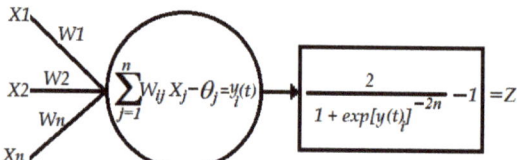

Figure 4. Representation of an artificial neuron called a perceptron.

The result of the transfer function is the output of the perceptron. It can be summarised as follows:

$$Z = \frac{2}{1+e^{-2\sum_{i=1}^{n} x_i w_i - \theta_i}} - 1. \tag{3}$$

The key characteristic of an ANN is its capacity to learn. The algorithm used to learn in an MLP is backpropagation (BP), which updates the synaptic weights according to the existing error between the value calculated by the network and the required one [82,84,85].

To find out the nonlinear connections between two groups of data, a neural network needs to be "trained" [86]. This is the reason why the input data, as well as the results that the analyst wants to obtain, are provided to the network. The network that repeatedly uses BP changes the weights (which have a random value at the beginning of the network) until it finds that a group of them that achieve the expected results. Once it has been trained, new data are provided to the network and it is tested to check the goodness of the group of weights. If it is not satisfactory, the weights are readjusted. When the network is tested and its efficiency is optimal, it is ready to work. BP is a generalisation of the Widrow-Hoff law in multilayer networks with non-linear transfer functions. BP allows the artificial neural network to be a universal approximator of functions. Biased networks, such as a sigmoid layer, and a linear output layer can work as approximators of any function with a specific number of discontinuities. BP is a gradient descent algorithm, meaning that the network weights are moved along the negative of the performance function's gradient. By just implementing the backpropagation learning, we can update the network weights and biases to make the performance function decrease quicker via the negative gradient.

BP is used to estimate the error between the output of an ANN and the goal. The procedure consists of proposing an error or cost function, which measures the network's performance. This function is determined by the synaptic weights (W). We can obtain the weight upgrade rule by means of the optimization methodology used in the error function. The error function is defined as E(W), which shows the mistake (E) that has been produced by the network. This error is converted into a cost function through the mean quadratic error.

The minimization of the cost function is done by means of a descent down the gradient in the hidden layer and the output layer. The upgrade of the weights is done by deriving the transfer functions.

The steps taken to train an MLP using BP are the following:

1. The weights and thresholds ($t = 0$) are randomly assigned.
2. For any pattern (μ) of input data:
   a. Execute the network to obtain the output for the μ pattern.
   b. Obtain the errors in hidden and output layers.
   c. Calculate the increase of weight and threshold for each μ pattern.
3. Calculate the total increase in all weights and the threshold for all patterns.
4. Upgrade the weights and thresholds.
5. Calculate the new error for $t = t + 1$ and return to step 2 [77].

This process is carried out for every learning set pattern. The upgrade of the weights and thresholds is done after the variation of weights for each pattern. After accumulating all these variations, all the weights are upgraded. This scheme is known as "batch learning".

The most common mistake made with ANNs is overtraining. The net learns so much that it fits exactly to input patterns. However, the problem is that, perhaps, this overtrained net will not be able to generalise and estimate future patterns. The solution is the early stop. We stop the training when we detect an increase in the total error.

All neural nets are MLP types with three layers: input, hidden and output. Following Demuth et al.'s criterion [87], both the input and hidden layers had the same number of neurons (nodes) as the number of the variables of the model. All the inputs were normalised according to their maximums and minimums in order to be able to train the network. The transfer function was a tan-sigmoid in all nodes in the hidden layer, while the function of the neurons in the output layer was linear. The output layer had only one neuron, which gave us the result of the neural net. We trained the neural nets with a backpropagation algorithm with an early stop to avoid overtraining. The aim was to obtain a better generalisation of the final model. The entire process was done using Matlab. Aside from neural networks, in this study, we estimated hedonic equations using ordinary least squares (OLS) (see [88,89]) and QR (we estimated 10th, 25th, 50th, 75th and 90th quantiles; Appendix A shows the quantile regression model used, which was based on [90]. In order to calculate the price characteristics (including time and location), the following equation was estimated [91]:

$$\text{Price}_{it} = \beta_0 + \sum_{k=1}^{k=K} \beta_k X_{ik} + \sum_{l=1}^{l=L} \alpha_l D_l + \sum_{t=1}^{t=T} \delta_t D_t + e_{it} \quad (4)$$

where the aim was to try to explain the price of a dwelling (Price_{it}) based on its characteristics (X_{ik}), the postal code in which it is located (D_l) and the year (D_t) in order to know the time trend. Finally, β_k, α_l and δ_t are parameters and e_{it} is the disturbance term, which follows the usual assumptions: the disturbance term is distributed as a normal function and is not correlated, and though it presents heteroskedasticity, the variance of the errors has been estimated in a robust way.

Therefore, this regression model provided estimates of the homogeneous parameters of dwellings, and the hedonic price theory justified its application. In the context of housing, it can be easily appreciated that the valuations that individuals make in relation to the physical characteristics of their dwellings differ according to their prices. Therefore, we aimed to find out the behaviour of the explanatory variables, as well as the price distribution. Consequently, an estimator that allows for heterogeneous responses was required: the estimator stemming from the QR (β_i). Additionally, a median-based (quantile) estimator was also appealing, given that it is less sensitive to outliers than a mean-based estimator. Thus, the bias from unobserved characteristics (i.e., renovation, quality) should be smaller.

In an estimated QR, before the estimation, the target is a parameter that is specified. On the one hand, let e_{it} be the residual implied by the econometric model (Equation (4)). On the other hand, let q represent the target quantile from the distribution residuals. Thus, the quantile parameter estimates are the coefficients that minimise the following objective function:

$$\sum_{e_{it}>0} 2q|e_{it}| + \sum_{e_{it}<0} 2(1-q)|e_{it}|. \quad (5)$$

For instance, equal weights are given to positive and negative residuals at the median ($q = 0.5$). However, at the 90th percentile ($q = 0.9$), more weight is given to positive residuals. Then, Equation (5) will be minimized at a set of parameter values; where 100q% of the residuals are positive. In this regard, this criterion is classically known as minimum absolute deviations. As a matter of fact, it tends to be used by employing the Koenker and Bassett Jr. [92] algorithm.

As far as the performances of the models were concerned, we used the following:

- Mean squared error (MSE): the mean of the square distance between the target value and the estimated value. It measures the quality of the estimator by measuring the mean squared error of our estimations. The higher this value is, the worse the model is.

$$\text{MSE}(y, \hat{y}) = \frac{1}{n} \sum_{i=0}^{n} (y_i - \hat{y}_i)^2 \qquad (6)$$

- Root mean squared error (RMSE): the square root of the MSE, i.e., it is calculated as the square root of the average of the quadratic differences between a variable and its estimation. RMSE is a measure of accuracy. It measures the amount of error between two datasets. To put it another way, it compares an estimated value and a known or observed value. This is one of the most commonly used statistics.

$$\text{RMSE}(y, \hat{y}) = \sqrt{\frac{\sum_{i=1}^{n} (y_i - \hat{y}_i)^2}{N}} \qquad (7)$$

- Mean absolute error (MAE): the mean absolute distance between the target value and the estimated value, i.e., the average of the sum of the absolute differences between a variable and its estimation. The same scale as the data being measured is used in MAE. It is known as a scale-dependent measure of accuracy and, thus, it cannot be used to make comparisons between series using different scales.

$$\text{MAE}(y, \hat{y}) = \frac{1}{n} \sum_{i=0}^{n} |y_i - \hat{y}_i| \qquad (8)$$

- Mean absolute percentage error (MAPE): the mean absolute distance between the target value and the estimated value divided by the target value, i.e., the average sum of the relative difference between a variable and its estimation. It is a measure of the estimation's accuracy. The mean absolute percentage error is an indicator of the performance of the demand estimation, which measures the size of the absolute error in percentage terms. It is useful even when the volume of demand for the product is not known since it is a relative measure.

$$\text{MAPE}(y, \hat{y}) = \frac{1}{n} \sum_{i=1}^{n} \left| \frac{y_i - \hat{y}_i}{y_i} \right| \qquad (9)$$

- R-squared coefficient (R^2): this provides information regarding to what extent the variance of a variable explains the variance of another variable. It is calculated as one minus the proportion between the square error from an estimation of a variable and the square error from the average of the same variable. It provides the measure of the accuracy of replication. The R^2 is the indicator that allowed us to know how well these results can be estimated. Therefore, R^2 is the variation percentage of the response variable that explains its relationship with one or more predictor variables. It can be said that, generally, the higher R^2 is, the better the model fits the data.

$$R^2(y, \hat{y}) = 1 - \frac{\sum_{i=0}^{n} (y_i - \hat{y}_i)^2}{\sum_{i=0}^{n} (y_i - \bar{y})^2} \qquad (10)$$

3. Data

Two datasets consisting exclusively of properties located in Catalonia were analysed. Dataset 1 was provided by a Spanish savings bank that was the result of a merger of three savings banks. The 163,871 properties included in dataset 1 were valued from 1994 to 2010 by independent appraisal companies. Dataset 2 was provided by a former Spanish savings bank that was also the result of a merger of three savings banks. Nevertheless, when the database was provided, the savings bank had been transformed into a commercial bank. The 24,781 properties included in dataset 2 were valued from 2004 to 2013 by independent appraisal companies. As explanatory variables, nine hedonic variables (dwelling characteristics), among which the postal code and year of the observation were included. Table 1 shows these variables and their definitions in detail, while Tables 2 and 3 show the descriptive statistics for quantitative, as well as qualitative and dichotomous variables, respectively.

Table 1. Definitions of the explanatory variables.

	Variable	Type	Definition
	ln(Price$_A$)	Quantitative	Natural logarithm of the appraisal price
	Height	Qualitative	The height of the house ranging from −2 to 19
	Elevator	Dichotomous	Whether the access to the house is by means of an elevator (1 = yes, 0 = no)
	Heating	Dichotomous	Whether the house uses a heating system (1 = yes, 0 = no)
Hedonic variables	Pool	Dichotomous	Whether the house or the residents' association property includes a swimming pool (1 = yes, 0 = no)
	Gardens	Dichotomous	Whether the house or the residents' association property includes a garden (1 = yes, 0 = no)
	Size	Quantitative	Constructed area of the house in square meters
	Condition	Dichotomous	Physical state of the house (meaning 1 = good, 0 = bad)
	Baths	Quantitative	Number of baths per house
	Rooms	Quantitative	Number of rooms per house
	PC	Qualitative	Postal code
	Year	Quantitative	Year when the house was priced

Table 2. Descriptive statistics of the quantitative variables.

Variable	Dataset 1		Dataset 2	
	Mean	Std. Dev.	Mean	Std. Dev.
ln(Price$_A$)	12.025	0.683	11.839	0.553
Size	131.025	80.575	85.672	63.464
Baths	1.595	0.703	1.295	0.660
Rooms	3.127	0.855	2.533	1.233
N	163,871		24,781	

Table 3. Descriptive statistics of the qualitative and dichotomous variables.

Variable	Dataset 1		Dataset 2	
	Median	Mode	Median	Mode
Height	2.000	0.000	1.000	0.000
Elevator	0.000	0.000	0.000	0.000
Heating	1.000	1.000	0.000	0.000
Pool	0.000	0.000	0.000	0.000
Gardens	0.000	0.000	0.000	0.000
Condition	1.000	1.000	0.000	0.000
N	163,871		24,781	

The combined number of properties analysed was 188,652, which represented 4.88% of the total number of the Catalan housing stock according to the official statistics published by INE [71]. On the other hand, the number of postal codes analysed amounted to 632 in dataset 1 and 607 in dataset 2; this meant that the properties analysed were not concentrated in a specific area, on the contrary, they are a good sample of Catalan housing (Catalonia comprises 1146 postal codes). Furthermore, the years analysed—1994 to 2013—covered both the rise and fall of real estate prices in Catalonia. Finally, according to the official statistics published by the Spanish Government by means of Ministry of Transport, Mobility and Urban Agenda (Ministerio de Transportes, Movilidad y Agenda Urbana) [93], the number of housing transactions from 2004 to 2013 was 890,554. On the other hand, the number of properties analysed in which prices corresponded to these years totalled 122,179. Hence, this amounted to 13.72% of the whole Catalan housing market for the decade studied.

4. Results and Discussion

Twelve models were created per dataset using the natural logarithm of the appraisal as the variable to be explained: four ANNs, four SLRs and four QRs. We have used some quasi-Newton algorithms (such as the Broyden-Fletcher-Goldfarb-Shanno (BFGS) method [87] and one-step secant), with the Levenberg-Marquardt algorithm being the one that performed better. Therefore, we have used supervised ANNs, with backpropagation (Levenberg-Marquardt) learning algorithms and an early stop. Table A1 in Appendix B shows the architecture of the developed ANN model. The models differ because different explanatory variables were used to create them: 1 means that only hedonic variables were used; 2 means that hedonic variables and postal code were used; 3 means that hedonic variables, postal code and year were used, therefore, all the explanatory variables were used; 4 means that all the explanatory variables were used and that we also controlled for postal code and year by means of transforming them into dummy variables.

The performances of these models are presented in Table 4 for dataset 1 and in Table 5 for dataset 2 in terms of MSE, RMSE, MAE, MAPE and R^2. We tested the models by means of a dataset of properties with transaction prices instead of appraisal prices, obtaining similar results (see Table A2).

Table 4. Comparison of the performances of artificial neural networks (ANNs), semi-log regressions (SLRs) and quantile regressions (QRs) for dataset 1.

Performance Measure	ANN 1	ANN 2	ANN 3	ANN 4	SLR 1	SLR 2	SLR 3	SLR 4	QR 1	QR 2	QR 3	QR 4
MSE	0.2933	0.2750	0.1107	0.1096	0.3141	0.1622	0.1162	0.1162	0.8533	0.8332	0.3120	0.2994
RMSE	0.5416	0.5244	0.3326	0.3311	0.5604	0.4028	0.3409	0.3409	0.9237	0.9128	0.5586	0.5472
MAE	0.4273	0.4128	0.2269	0.2278	0.4458	0.3982	0.1983	0.1969	0.7911	0.7808	0.4763	0.4642
MAPE	0.0361	0.0349	0.0192	0.0193	0.0371	0.0328	0.0188	0.0186	0.0772	0.0667	0.0432	0.0398
R^2	0.3712	0.4105	0.7628	0.7651	0.3267	0.4680	0.8180	0.8200	0.2358	0.2452	0.4084	0.4124

Table 5. Comparison of the performance of the ANNs, SLRs and QRs for dataset 2.

Performance Measure	ANN 1	ANN 2	ANN 3	ANN 4	SLR 1	SLR 2	SLR 3	SLR 4	QR 1	QR 2	QR 3	QR 4
MSE	0.2088	0.1706	0.1238	0.1273	0.2294	0.1190	0.0866	0.0853	0.6363	0.5491	0.4900	0.4742
RMSE	0.4569	0.4131	0.3519	0.3568	0.4790	0.3450	0.2943	0.2920	0.7977	0.7410	0.7000	0.6886
MAE	0.3571	0.3196	0.2595	0.2633	0.3730	0.2711	0.2172	0.2159	0.6851	0.6311	0.5022	0.4979
MAPE	0.0305	0.0272	0.0221	0.0224	0.0321	0.0218	0.0181	0.0175	0.0580	0.0550	0.0455	0.0445
R^2	0.3171	0.4419	0.5951	0.5835	0.2492	0.5900	0.7081	0.7150	0.2129	0.2470	0.5351	0.5360

The results suggest that the ANNs and SLRs were better tools than QRs for modelling housing prices in Catalonia. In fact, the results in terms of all common performance measures for all the models and datasets were better for the ANNs and SLRs than for the QRs. On the one hand, the ANNs were better than SLRs when only hedonic variables were used. On the other hand, when more variables were used, SLRs obtained better results using dataset 2, whereas the performance results were not conclusive for dataset 1 when more variables were used, independent of whether the year was considered a dummy variable. Finally, all the models obtained better results when more variables were included and the use of time as a dummy variable slightly enhanced the results obtained for the SLRs and QRs for all datasets. The improvement of the results from ANN1 to ANN2 was studied by McGreal et al. [48], who asserted that better results were obtained by neural networks when the postal code was used as a delimiter. This makes sense because location effects are crucial when estimating real estate prices. Following the same line of reasoning, the real estate market is dynamic and time-fixed effects are also crucial in their estimation. On the contrary, the use of time as a dummy variable does not improve the models obtained by means of ANN. In other words, when using an ANN, transforming a quantitative variable into dummies will generate the same results. This was confirmed by Peterson and Flanagan [45], who stated that since the parametric estimation in an ANN does not depend on the range of the regressive matrix, ANNs are better than the models that need to use large sets of dummies. In fact, the performance results for ANN models were worse for dataset 2 and inconclusive for dataset 1 when time was used as a dummy variable in comparison to when it was considered as a quantitative one.

The results suggest that the ANN models improved when the analysed dataset was larger. In fact, when the smallest dataset was used, the SLR results were better than the ones obtained by the ANNs. Therefore, we agree with Worzala et al. [49], who compared ANNs with traditional multiple regression models and no evidence was found demonstrating that ANNs are superior for valuation analysis. Nevertheless, our results demonstrate that they are not worse, except regarding the R^2 coefficient, where only similar results were obtained using SLR methodology when the largest dataset is used. This is confirmed by Nghiep and Cripps [50], who determined that ANNs obtain results that are similar to those obtained using regression models when a large sample is used.

5. Conclusions

This paper presents new evidence to compare the performances of QR and SLR hedonic models relative to ANNs using data of properties that belonged to two banks. The aim of this study was threefold:

First, this study aimed to cover the literature gap in regard to the comparison of QRs and ANNs for assessing hedonic prices in housing, with this being the first article to include this comparison. The results suggest that QRs are worse tools than ANNs when modelling housing prices in Catalonia. Therefore, QR valuation modelling cannot be considered as an alternative to ANNs given that its performance was worse for all datasets, regardless of the number of variables used.

Second, when using all the variables, the SLRs performed better than the ANNs with the smallest dataset, whereas the results were not conclusive in regard to the largest dataset. Therefore, in the specific case of Catalonia, we cannot confirm the fact observed in other markets that suggest that ANNs perform better than SLR when assessing real estate. Third, out of the three models analysed and according to the results obtained, Spanish banks should use a model for housing mass appraisals that matches their size. Small and medium banks (mainly cooperatives and savings banks) should use SLRs rather than ANNs given that SLRs are better when the dataset was smaller. On the other hand, large banks (mainly commercial banks) can use either SLRs or ANNs given that their performance was similar for larger datasets. Finding out the optimal way to value properties registered in banks' balance sheets has been one of the greatest challenges that the banking industry has faced in recent years. An optimal valuation offers two advantages: first, the real financial situation of the bank is established; second, if the property is valued according to the market, it can be sold more quickly and the revenues obtained will be maximised. Overall, given that this study was carried out with data obtained previous to the recent legislation that has limited rental prices in Catalonia (Law 11/2020, issued on 18 September 2020) and that the

Catalan real estate market is representative of the Spanish market, the conclusions can be generalised to the whole country of Spain.

With regard to the limitations of this paper, we must recognise that the main one has to do with the fact that datasets used included prices through to 2013. It would be very useful to obtain more recent and similar databases. Nonetheless, it is highly unlikely that the authors may obtain such a database in the future since it is usually not available to researchers due to opacity of information given by banks. Nevertheless, the conclusions obtained by means of this study can be considered important because the lifespan of the data analysed ranged from 1994 to 2013. Therefore, boom and recession years in the housing industry were included due to the fact that this time horizon encompassed the rise and fall of the Spanish real estate bubble.

Future lines of research could include the analysis of more recent large databases, in the event of them becoming available. Additionally, by means of a simulation exercise, future studies could analyse the extent to which banks would have benefited—by means of an increase of revenues, capital gains generation and the reversal of impairment losses—from having used these models.

Author Contributions: Conceptualization, J.T.-P., P.G.-E. and C.P.-R.; data curation, J.T.-P.; formal analysis, J.T.-P. and P.G.-E.; funding acquisition, C.P.-R.; investigation, J.T.-P., P.G.-E. and C.P.-R.; methodology, J.T.-P. and P.G.-E.; project administration, J.T.-P.; resources, J.T.-P. and P.G.-E.; software, J.T.-P. and P.G.-E.; supervision, J.T.-P.; validation, J.T.-P. and P.G.-E.; visualization, J.T.-P. and P.G.-E.; writing—original draft, J.T.-P., P.G.-E. and C.P.-R.; writing—review & editing J.T.-P., P.G.-E. and C.P.-R. All authors have read and agreed to the published version of the manuscript.

Funding: We acknowledge the support of the publication fee provided by Universidad Rey Juan Carlos.

Institutional Review Board Statement: Not applicable.

Informed Consent Statement: Not applicable.

Data Availability Statement: Restrictions apply to the availability of these data. The data are not publicly available due to privacy issues and cannot be shared.

Conflicts of Interest: The authors declare no conflict of interest.

Appendix A

Following [90], the quantile regression model used is as follows. Let (y_i, x_i), $i = 1, \ldots, n$, be a sample from some population, where x_i is a $K \times 1$ vector of regressors. It was assumed that:

$$y_i = x'_i \beta_\theta + u_{\theta_i}, \quad Quant_\theta(y_i | x_i) = x'_i \beta_\theta, \tag{A1}$$

where $Quant_\theta(y_i | x_i)$ represents the conditional quantile of y_i, which is conditional on the x_i regressor vector. If $F_{\mu_\theta}(\cdot)$ is known, we can use different methods to estimate β_θ. However, u_{θ_i} as the distribution of the error term was not specified but it was assumed to satisfy the following quantile restriction: $Quant_\theta(u_{\theta_i} | x_i) = 0$.

In general, let $\hat{u}_\theta$ be the θth sample quantile ($0 < \theta < 1$) of y that solves:

$$\min_b \left\{ \sum_{i:y_i \geq b} \theta |y_i - b| + \sum_{i:y_i < b} (1 - \theta)|y_i - b| \right\}. \tag{A2}$$

Similarly, $\hat{\beta}_\theta$, the estimator for β_θ in (A1), which is termed the θth quantile regression, solves Equation (A3):

$$\min_\beta \frac{1}{n} \left\{ \sum_{i:y_i \geq x'_i \beta} \theta |y_i - x'_i \beta| + \sum_{i:y_i < x'_i \beta} (1 - \theta)|y_i - x'_i \beta| \right\} = \min_\beta \frac{1}{n} \sum_{i=1}^{n} \rho_\theta(u_{\theta_i}), \tag{A3}$$

where $\rho_\theta(\lambda) = (\theta - I(\lambda < 0))\lambda$ is the check function and $I()$ is the usual indicator function. Therefore, Equation (A3) can be written as follows:

$$\min_\beta \frac{1}{n} \sum_{i=1}^{n} \left(\theta - \frac{1}{2} + \frac{1}{2}\text{sgn}(y_i - x'_i b) \right)(y_i - x'_i b). \tag{A4}$$

Equation (A5) gives the $K \times 1$ vector of first-order conditions for Equation (A4):

$$\frac{1}{n} \sum_{i=1}^{n} \left(\theta - \frac{1}{2} + \frac{1}{2}\text{sgn}(y_i - x'_i \hat{\beta}_\theta) \right) x_i = 0. \tag{A5}$$

As a matter of fact, the specified first-order conditions in Equation (A5) implies a moment function that fits into the generalised methods of moments framework. The moment function is defined as follows:

$$\psi(x_i, y_i, \beta) = \left(\theta - \frac{1}{2} + \frac{1}{2}\text{sgn}(y_i - x'_i \hat{\beta}_\theta) \right) x_i. \tag{A6}$$

The validity of $\psi(\cdot)$ in Equation (A6) as a moment function is established by the fact that under certain regularity conditions, $E[\psi(x_i, y_i, \beta_\theta)] = 0$. Thus, the generalised method of moments framework can be applied to establish the asymptotic normality and consistency of $\hat{\beta}_\theta$. Specifically, it can be shown, under certain regularity conditions (see [90]), that:

$$\sqrt{n}(\hat{\beta}_\theta - \beta_\theta) \xrightarrow{L} N(0, \Lambda_\theta) \tag{A7}$$

where:

$$\Lambda_\theta = \theta(1-\theta)\left(E[f_{u_\theta}(0|x_i)x_i x'_i]\right)^{-1} E[x_i x'_i]\left(E[f_{u_\theta}(0|x_i)x_i x'_i]\right)^{-1}. \tag{A8}$$

If $f_{u_\theta}(0|x) = f_{u_\theta}(0)$ with probability 1, then Λ_θ in Equation (A8) can be simplified to:

$$\Lambda_\theta = \frac{\theta(1-\theta)}{f^2_{u_\theta}(0)} \left(E[x_i x'_i]\right)^{-1}. \tag{A9}$$

Appendix B

Table A1. Architecture of the developed ANN model.

Property	Structure
Number of hidden neurons	The same number as the input layer
Transfer function in the hidden layer	Tan-sigmoid
Transfer function in the neuron of output layer	Linear
Type of learning rule	Backpropagation with the Levenberg–Marquardt algorithm
Control of overlearning	Early stop

Table A2. Comparison of the performance of the ANNs, SLRs and QRs for the dataset with transaction prices.

Performance Measure	ANN 1	ANN 2	ANN 3	ANN 4	SLR 1	SLR 2	SLR 3	SLR 4	QR 1	QR 2	QR 3	QR 4
MSE	0.1666	0.1334	0.1056	0.1675	0.2221	0.1416	0.0671	0.0663	0.5646	0.4756	0.4170	0.4072
RMSE	0.4082	0.3652	0.3249	0.4092	0.4345	0.3469	0.2389	0.2374	0.6927	0.6358	0.5954	0.5883
MAE	0.3187	0.2819	0.2436	0.3163	0.3391	0.2758	0.1903	0.1721	0.5769	0.5436	0.4984	0.4875
MAPE	0.0268	0.0237	0.0204	0.0274	0.0297	0.0199	0.0163	0.0148	0.0503	0.0480	0.0430	0.0423
R^2	0.3072	0.4449	0.5609	0.4934	0.1783	0.4161	0.6436	0.6562	0.1429	0.1946	0.2269	0.2278

References

1. Akin, O.; Montalvo, J.G.; Villar, J.G.; Peydró, J.-L.; Raya, J.M. The real estate and credit bubble: Evidence from Spain. *SERIEs* **2014**, *5*, 223–243. [CrossRef]
2. San-José, L.; Retolaza, J.L.; Torres-Pruñonosa, J. Efficiency in Spanish banking: A multi-stakeholder approach analysis. Journal of International Financial Markets. *Instit. Money* **2014**, *32*, 240–255. [CrossRef]
3. San-José, L.; Retolaza, J.L.; Torres-Pruñonosa, J. Eficiencia social en las cajas de ahorro españolas transformadas en bancos [Social Efficiency in Savings Banks Transformed into Commercial Banks in Spain]. *Trimest Econ.* **2020**, *87*, 759–787. [CrossRef]
4. Griliches, Z. *Price Indexes and Quality Change*; Harvard University Press: Cambridge, MA, USA, 1971.
5. Court, L.M. Entrepreneurial and consumer demand theories for commodity spectra. *Econometrica* **1941**, *9*, 135–162. [CrossRef]
6. Tinbergen, J. Some remarks on the distribution of labour incomes. *Int. Econ. Pap.* **1951**, *1*, 195–207.
7. Rosen, S. Hedonic Prices and Implicit Markets: Product Differentiation in Pure Competition. *J. Politi Econ.* **1974**, *82*, 34–55. [CrossRef]
8. Lancaster, K.T. A New Approach to Consumer Theory. *J. Political Econ.* **1966**, *74*, 132–157. [CrossRef]
9. Bartik, T.J. The Estimation of Demand Parameters in Hedonic Price Models. *J. Politi Econ.* **1987**, *95*, 81–88. [CrossRef]
10. Bin, O. A semiparametric hedonic model for valuing wetlands. *Appl. Econ. Lett.* **2005**, *12*, 597–601. [CrossRef]
11. Bover, O.; Velilla, P. Hedonic house prices without characteristics: The case of new multiunit housing. In *ECB Working Paper 117*; European Central Bank: Frankfurt, Germany, 2002.
12. Garcia, J.; Raya, J.M. Price and Income Elasticities of Demand for Housing Characteristics in the City of Barcelona. *Reg. Stud.* **2011**, *45*, 597–608. [CrossRef]
13. Mendelsohn, R. Estimating the Structural Equations of Implicit Markets and Household Production Functions. *Rev. Econ. Stat.* **1984**, *66*, 673–677. [CrossRef]
14. Mills, E.S.; Simenauer, R. New Hedonic Estimates of Regional Constant Quality House Prices. *J. Urban Econ.* **1996**, *39*, 209–215. [CrossRef]
15. Palmquist, R.B. Estimating the Demand for the Characteristics of Housing. *Rev. Econ. Stat.* **1984**, *66*, 394–404. [CrossRef]
16. Kuminoff, N.V.; Parmeter, C.F.; Pope, J.C. Which hedonic models can we trust to recover the marginal willingness to pay for environmental amenities? *J. Environ. Econ. Manag.* **2010**, *60*, 145–160. [CrossRef]
17. Li, H.; Wei, Y.D.; Yu, Z.; Tian, G. Amenity, accessibility and housing values in metropolitan USA: A study of Salt Lake County, Utah. *Cities* **2016**, *59*, 113–125. [CrossRef]
18. Li, H.; Wei, Y.D.; Wu, Y.; Tian, G. Analyzing housing prices in Shanghai with open data: Amenity, accessibility and urban structure. *Cities* **2019**, *91*, 165–179. [CrossRef]
19. Bruegge, C.; Carrión-Flores, C.; Pope, J.C. Does the housing market value energy efficient homes? Evidence from the energy star program. *Reg. Sci. Urban Econ.* **2016**, *57*, 63–76. [CrossRef]
20. Wu, C.; Ye, X.; Du, Q.; Luo, P. Spatial effects of accessibility to parks on housing prices in Shenzhen, China. *Habitat Int.* **2017**, *63*, 45–54. [CrossRef]
21. Raya, J.M.; García-Estévez, P.; Prado-Román, C.; Torres-Pruñonosa, J. Living in a smart city affects the value of a dwelling? In *Sustainable Smart Cities: Creating Spaces for Technological, Social and Business Development Innovation, Technology, and Knowledge Management*; Peris-Ortiz, M., Bennett, D., Yábar, D.P.-B., Eds.; Springer: Berlin/Heidelberg, Germany, 2017; pp. 193–198.
22. Pérez-Sánchez, V.R.; Mora-García, R.T.; Pérez-Sánchez, J.C.; Céspedes-López, M.F. La influencia de las caracte-rísticas de las viviendas de segunda mano en sus precios de venta: Evidencias en el mercado alicantino. *Infor. Constr.* **2020**, *72*, e345. [CrossRef]
23. Coulson, N.E.; McMillen, D.P. The Dynamics of Intraurban Quantile House Price Indexes. *Urban Stud.* **2007**, *44*, 1517–1537. [CrossRef]
24. García, J.; Raya, J.M. Use of a Gini index to examine housing price heterogeneity: A quantile approach. *J. Hous. Econ.* **2015**, *29*, 59–71.
25. McMillen, D.P. Changes in the distribution of house prices over time: Structural characteristics, neighborhood, or coefficients? *J. Urban Econ.* **2008**, *64*, 573–589. [CrossRef]
26. McMillen, D.P.; Thorsnes, P. Housing Renovations and the Quantile Repeat-Sales Price Index. *Real. Estate Econ.* **2006**, *34*, 567–584. [CrossRef]
27. Nicodemo, C.; Raya, J.M. Change in the distribution of house prices across Spanish cities. *Reg. Sci. Urban Econ.* **2012**, *42*, 739–748. [CrossRef]
28. Deng, Y.; McMillen, D.P.; Sing, T.F. Private residential price indices in Singapore: A matching approach. *Reg. Sci. Urban Econ.* **2012**, *42*, 485–494. [CrossRef]
29. Liao, W.; Wang, X. Hedonic house prices and spatial quantile regression. *J. Hous. Econ.* **2012**, *21*, 16–27. [CrossRef]
30. Kholodilin, K.A.; Ulbricht, D. Urban House Prices: A Tale of 48 Cities. *Econ. Open-Access E-J.* **2015**, *9*, 1–43. [CrossRef]
31. Waltl, S.R. Variation Across Price Segments and Locations: A Comprehensive Quantile Regression Analysis of the Sydney Housing Market. *Real Estate Econ.* **2016**, *47*, 723–756. [CrossRef]
32. Zhang, L.; Yi, Y. What contributes to the rising house prices in Beijing? A decomposition approach. *J. Hous. Econ.* **2018**, *41*, 72–84. [CrossRef]
33. Peng, C.-W.; Tsai, I.-C. The long- and short-run influences of housing prices on migration. *Cities* **2019**, *93*, 253–262. [CrossRef]

34. Mora-Garcia, R.-T.; Cespedes-Lopez, M.-F.; Perez-Sanchez, V.R.; Marti, P.; Perez-Sanchez, J.-C. Determinants of the Price of Housing in the Province of Alicante (Spain): Analysis Using Quantile Regression. *Sustainability* **2019**, *11*, 437. [CrossRef]
35. Chien, M.-S.; Setyowati, N. The effects of uncertainty shocks on global housing markets. *Int. J. Hous. Mark. Anal.* **2020**, *14*, 218–242. [CrossRef]
36. McMillen, D.; Shimizu, C. Decompositions of house price distributions over time: The rise and fall of Tokyo house prices. *Real. Estate Econ.* **2020**. [CrossRef]
37. Ekeland, I.; Heckman, J.J.; Nesheim, L. Identifying Hedonic Models. *Am. Econ. Rev.* **2002**, *92*, 304–309. [CrossRef]
38. Selim, H. Determinants of house prices in Turkey: Hedonic regression versus artificial neural network. *Expert Syst. Appl.* **2009**, *36*, 2843–2852. [CrossRef]
39. White, H. Economic prediction using neural networks: The case of IBM daily stock returns. In Proceedings of the IEEE International Conference on Neural Networks, San Diego, CA, USA, 24–27 June 1988; pp. 451–459.
40. Din, A.; Hoesli, M.; Bender, A. Environmental Variables and Real Estate Prices. *Urban Stud.* **2001**, *38*, 1989–2000. [CrossRef]
41. Do, A.Q.; Grudnitski, G. A neural network approach to residential property appraisal. *Real Estate Apprais.* **1992**, *58*, 38–45.
42. Kauko, T. On current neural network applications involving spatial modelling of property prices. *Neth. J. Hous. Environ. Res.* **2003**, *18*, 159–181. [CrossRef]
43. Landajo, M.; Bilbao, C.; Bilbao, A. Nonparametric neural network modeling of hedonic prices in the housing market. *Empir. Econ.* **2011**, *42*, 987–1009. [CrossRef]
44. Limsombunchai, V.; Gan, C.; Lee, M. House Price Prediction: Hedonic Price Model vs. Artificial Neural Network. *Am. J. Appl. Sci.* **2004**, *1*, 193–201. [CrossRef]
45. Peterson, S.; Flanagan, A. Neural Network Hedonic Pricing Models in Mass Real Estate Appraisal. *J. Real Estate Res.* **2009**, *31*, 147–164. [CrossRef]
46. Tay, D.P.; Ho, D.K. Artificial Intelligence and the Mass Appraisal of Residential Apartments. *J. Prop. Valuat. Invest.* **1992**, *10*, 525–540. [CrossRef]
47. Curry, B.; Morgan, P.; Silver, M. Neural networks and non-linear statistical methods: An application to the modelling of price–quality relationships. *Comput. Oper. Res.* **2002**, *29*, 951–969. [CrossRef]
48. McGreal, S.; Adair, A.; McBurney, D.; Patterson, D. Neural networks: The prediction of residential values. *J. Prop. Valuat. Invest.* **1998**, *16*, 57–70. [CrossRef]
49. Worzala, E.; Lenk, M.; Silva, A. An Exploration of Neural Networks and Its Application to Real Estate Valuation. *J. Real Estate Res.* **1995**, *10*, 185–201. [CrossRef]
50. Nghiep, N.; Cripps, A. Predicting Housing Value: A Comparison of Multiple Regression Analysis and Artificial Neural Networks. *J. Real Estate Res.* **2001**, *3*, 313–336. [CrossRef]
51. Liu, J.-G.; Zhang, X.-L.; Wu, W.-P. Application of Fuzzy Neural Network for Real Estate Prediction. In *International Symposium on Neural Networks*; Springer: Berlin/Heidelberg, Germany, 2016; pp. 1187–1191.
52. Abidoye, R.B.; Chan, A.P.C. Improving property valuation accuracy: A comparison of hedonic pricing model and artificial neural network. *Pac. Rim Prop. Res. J.* **2017**, *24*, 71–83. [CrossRef]
53. Štubňová, M.; Urbaníková, M.; Hudáková, J.; Papcunová, V. Estimation of Residential Property Market Price: Comparison of Artificial Neural Networks and Hedonic Pricing Model. *Emerg. Sci. J.* **2020**, *4*, 530–538. [CrossRef]
54. Mayer, M.; Bourassa, S.C.; Hoesli, M.; Scognamiglio, D. Estimation and updating methods for hedonic valuation. *J. Eur. Real Estate Res.* **2019**, *12*, 134–150. [CrossRef]
55. Jiang, C.; Jiang, M.; Xu, Q.; Huang, X. Expectile regression neural network model with applications. *Neurocomputing* **2017**, *247*, 73–86. [CrossRef]
56. Xu, C.; Chen, H. A hybrid data mining approach for anomaly detection and evaluation in residential buildings energy data. *Energy Build.* **2020**, *215*, 109864. [CrossRef]
57. Tyrväinen, L.; Miettinen, A. Property Prices and Urban Forest Amenities. *J. Environ. Econ. Manag.* **2000**, *39*, 205–223. [CrossRef]
58. Geoghegan, J. The value of open spaces in residential land use. *Land Use Policy* **2002**, *19*, 91–98. [CrossRef]
59. Cropper, M.L.; Deck, L.B.; McConnell, K.E. On the Choice of Funtional Form for Hedonic Price Functions. *Rev. Econ. Stat.* **1988**, *70*, 668–675. [CrossRef]
60. Tabales, J.N.; Ocerín, J.M.C.; Carmona, F.R. Artificial Neural Networks for Predicting Real Estate Prices. *Rev. Métodos Cuantitativos Econ. Empresa* **2013**, *15*, 29–44.
61. Tabales, J.N.; Carmona, F.R.; Ocerín, J.M.C. Precios implícitos en valoración inmobiliaria urbana. *Rev. Constr.* **2013**, *12*, 116–126.
62. Tabales, J.M.N.; Carmona, F.J.R.; Ocerin, J.M.C.Y. Redes neuronales (RN) aplicadas a la valoración de locales comerciales. *Infor. Constr.* **2017**, *69*, 179. [CrossRef]
63. Baldominos, A.; Blanco, I.; Moreno, A.J.; Iturrarte, R.; Bernardez, O.; Afonso, C. Identifying Real Estate Opportunities Using Machine Learning. *Appl. Sci.* **2018**, *8*, 2321. [CrossRef]
64. Edelstein, R.H.; Quan, D.C. How Does Appraisal Smoothing Bias Real Estate Returns Measurement? *J. Real Estate Financ. Econ.* **2006**, *32*, 41–60. [CrossRef]
65. García-Montalvo, J.; Raya, J.M. Constraints on LTV as a Macroprudential Tool: A Precautionary Tale. *Oxf. Econ. Pap.* **2018**, *70*, 821–845. [CrossRef]

66. Wang, D.; Li, V.J. Mass Appraisal Models of Real Estate in the 21st Century: A Systematic Literature Review. *Sustainability* **2019**, *11*, 7006. [CrossRef]
67. Torres-Pruñonosa, J.; Retolaza, J.L.; San-José, L. Gobernanza multifiduciaria de stakeholders: Análisis comparado de la eficiencia de bancos y cajas de ahorros. *Revesco. Rev. Estud. Coop.* **2012**, *108*, 152–172. [CrossRef]
68. San-José, L.; Retolaza, J.L.; Torres-Pruñonosa, J. Empirical evidence of Spanish banking efficiency: The stakeholder theory perspective. In *Soft Computing in Management and Business Economics Studies in Fuzziness and Soft Computing*; Gil-Lafuente, A.M., Gil-Lafuente, J., Merigó-Lindahl, J.M., Eds.; Springer: Berlin/Heidelberg, Germany, 2012; pp. 153–165.
69. de La Paz, P.T.; Gabrielli, L. Housing Supply and Price Reactions: A Comparison Approach to Spanish and Italian Markets. *Hous. Stud.* **2015**, *30*, 1036–1063. [CrossRef]
70. Dol, K.; Mazo, E.C.; Llop, N.L.; Hoekstra, J.; Fuentes, G.C.; Etxarri, A.E. Regionalization of housing policies? An exploratory study of Andalusia, Catalonia and the Basque Country. *Neth. J. Hous. Environ. Res.* **2016**, *32*, 581–598. [CrossRef] [PubMed]
71. Instituto Nacional de Estadística. Available online: https://www.ine.es/ (accessed on 27 February 2021).
72. Ministerio de Agricultura, Alimentación y Medio Ambiente del Gobierno de España. Tercer Inventario Forestal Nacional. Available online: https://www.miteco.gob.es/es/biodiversidad/servicios/banco-datos-naturaleza/informacion-disponible/ifn3.aspx (accessed on 27 February 2021).
73. Instituto Geográfico Nacional. Available online: https://www.ign.es/web/ign/portal/inicio (accessed on 27 February 2021).
74. Ministerio de Transportes, Movilidad y Agenda Urbana. Estimación de Precios de Suelo Urbano. Available online: https://www.fomento.gob.es/BE2/?nivel=2&orden=36000000 (accessed on 27 February 2021).
75. Funahasi, K.I. On the approximate realization of continuous mapping by neural networks. *Neural Netw.* **1989**, *3*, 183–192. [CrossRef]
76. Hornik, K.; Stinchcombe, M.; White, H. Multilayer feedforward networks are universal approximators. *Neural Netw.* **1989**, *2*, 359–366. [CrossRef]
77. del Brío, B.M.; Sanz, A. *Redes Neuronales y Sistemas Borrosos*; Ra–ma Editorial: Madrid, Spain, 1997.
78. Minsky, M.; Papert, S. *Perceptrons*; The MIT Press: Cambridge, UK, 1969; pp. 1–20.
79. Widrow, B.; Hoff, M.E. *Adaptive Switching Circuits*; Stanford Univ Ca Stanford Electronics Labs: Stanford, CA, USA, 1960.
80. Vesanto, J.; Alhoniemi, E. Clustering of the self-organizing map. *IEEE Trans. Neural Netw.* **2000**, *11*, 586–600. [CrossRef]
81. Rosenblatt, F. The perceptron: A probabilistic model for information storage and organization in the brain. *Psychol. Rev.* **1958**, *65*, 386–408. [CrossRef]
82. Hecht-Nielsen, R. Theory of the backpropagation neural network. In *Neural Networks for Perception*; Academic Press: Cambridge, MA, USA, 1989; pp. 593–605.
83. Sánchez-Serrano, J.R.; Alaminos, D.; García-Lagos, F.; Callejón-Gil, A.M. Predicting Audit Opinion in Consolidated Financial Statements with Artificial Neural Networks. *Mathematics* **2020**, *8*, 1288. [CrossRef]
84. Rumelhart, D.E.; Hinton, G.E.; Williams, R.J. Learning representations by back-propagating errors. *Nat. Cell Biol.* **1986**, *323*, 533–536. [CrossRef]
85. Li, Y.; Chen, W. A Comparative Performance Assessment of Ensemble Learning for Credit Scoring. *Mathematics* **2020**, *8*, 1756. [CrossRef]
86. Kohonen, T. Self-organized formation of topologically correct feature maps. *Biol. Cybern.* **1982**, *43*, 59–69. [CrossRef]
87. Demuth, H.; Beale, M.; Hagan, M. *Neural Network ToolBox TM 6. User's Guide*; The MathWorks, Inc.: Natick, MA, USA, 2009.
88. Wooldridge, J.M. *Introductory Econometrics. A Modern Approach*, 7th ed.; Cengage Learning: Boston, MA, USA, 2020.
89. Gujarati, D.N.; Porter, D.C. *Econometría*, 5th ed.; McGraw Hill: New York, NY, USA, 2010.
90. Buchinsky, M. Recent Advances in Quantile Regression Models: A Practical Guideline for Empirical Research. *J. Hum. Resour.* **1998**, *33*, 88–126. [CrossRef]
91. Vilchez, J.R. Destination and Seasonality Valuations: A Quantile Approach. *Tour. Econ.* **2013**, *19*, 835–853. [CrossRef]
92. Koenker, R.; Bassett, G., Jr. Regression quantiles. *Econometrica* **1978**, *1*, 33–50. [CrossRef]
93. Ministerio de Transportes, Movilidad y Agenda Urbana. Transacciones Inmobiliarias (Compraventa). Available online: https://www.fomento.gob.es/be2/?nivel=2&orden=34000000 (accessed on 27 February 2021).

Article

A Conceptual Probabilistic Framework for Annotation Aggregation of Citizen Science Data

Jesus Cerquides [1,*], Mehmet Oğuz Mülâyim [1,*], Jerónimo Hernández-González [2], Amudha Ravi Shankar [3] and Jose Luis Fernandez-Marquez [3]

1 Institut d'Investigació en Intel·ligència Artificial (IIIA), CSIC, 08193 Cerdanyola, Spain
2 Department de Matemàtiques, Universitat de Barcelona, 08007 Barcelona, Spain; jeronimo.hernandez@ub.edu
3 Citizen Cyberlab, CUI, University of Geneva, CH-1227 Geneva, Switzerland; Amudha.RaviShankar@unige.ch (A.R.S.); JoseLuis.Fernandez@unige.ch (J.L.F.-M.)
* Correspondence: cerquide@iiia.csic.es (J.C.); oguz@iiia.csic.es (M.O.M.); Tel.: +34-935809570 (ext. 223) (J.C.)

Abstract: Over the last decade, hundreds of thousands of volunteers have contributed to science by collecting or analyzing data. This public participation in science, also known as citizen science, has contributed to significant discoveries and led to publications in major scientific journals. However, little attention has been paid to data quality issues. In this work we argue that being able to determine the accuracy of data obtained by crowdsourcing is a fundamental question and we point out that, for many real-life scenarios, mathematical tools and processes for the evaluation of data quality are missing. We propose a probabilistic methodology for the evaluation of the accuracy of labeling data obtained by crowdsourcing in citizen science. The methodology builds on an abstract probabilistic graphical model formalism, which is shown to generalize some already existing label aggregation models. We show how to make practical use of the methodology through a comparison of data obtained from different citizen science communities analyzing the earthquake that took place in Albania in 2019.

Keywords: data quality; citizen science; consensus models

Citation: Cerquides, J.; Mülâyim, M.O.; Hernández-González, J.; Ravi Shankar, A.; Fernandez-Marquez, J.L. A Conceptual Probabilistic Framework for Annotation Aggregation of Citizen Science Data. *Mathematics* **2021**, *9*, 875. https://doi.org/10.3390/math9080875

Academic Editor: Snezhana Gocheva-Ilieva

Received: 26 February 2021
Accepted: 13 April 2021
Published: 15 April 2021

Publisher's Note: MDPI stays neutral with regard to jurisdictional claims in published maps and institutional affiliations.

Copyright: © 2021 by the authors. Licensee MDPI, Basel, Switzerland. This article is an open access article distributed under the terms and conditions of the Creative Commons Attribution (CC BY) license (https://creativecommons.org/licenses/by/4.0/).

1. Introduction

Citizen science (CS) is scientific research conducted, in whole or in part, by amateur (or nonprofessional) scientists [1]. Haklay [2] offers an overview of the typologies of the level of citizen participation in citizen science, which range from "crowdsourcing" (level 1), where the citizen acts as a sensor, to "distributed intelligence" (level 2), where the citizen acts as a basic interpreter, to "participatory science", where citizens contribute to problem definition and data collection (level 3), to "extreme citizen science", which involves collaboration between the citizen and scientists in problem definition, collection and data analysis. In this work we focus on distributed intelligence citizen science tasks where the citizen provides basic interpretations of data. Examples of tasks under the scope would be (i) citizens classifying images of living species in a taxonomy or (ii) citizens determining whether a tweet contains relevant information for the evaluation of a specific natural disaster.

The current practice tackles distributed intelligence citizen science tasks by (i) determining a set of tasks that need to be solved/annotated/interpreted; then (ii) distributing each of those tasks to a set of citizens who solve them; and finally (iii) aggregating the solutions from the citizens to obtain a consensus solution for each task. Steps (i) and (ii) are usually guided by tools such as Pybossa [3]. In this work we focus mainly on stage (iii), that is, how to perform the aggregation of different citizens' annotations for a specific task. However, we will see that models used to aggregate citizens' annotations can also be used to influence stage (ii) in the decision of how many citizens, and more specifically, which citizens, should be requested to solve a specific task.

Data fusion in citizen science is a very wide topic [4–6]. For the specific subtopic of interest in this paper, that of label aggregation, the state-of-the-art citizen science applications aggregate the annotations of different citizen scientists on a specific task by using majority voting [7]. That is, the option that gets the larger number of votes is the option considered as correct, with each citizen's opinion having the very same weight. However, the problem of aggregating different annotations has received a lot of attention from the statistics and machine learning communities, where more complex aggregation procedures have been introduced. The contributions of our work are (i) the introduction of a probabilistic model-based approach to aggregate annotations with citizen science applications in mind, and (ii) providing a case study that shows the added value in a citizen science scenario: the comparison of the data quality that can be obtained from different communities, where a gold set for measuring quality is unavailable. Furthermore, we approach the problem from an epistemic probabilistic perspective [8], relying as much as possible on information-theoretic concepts [9,10].

We start by reviewing related work in Section 2. After that, we provide an abstract probabilistic model for the annotation problem in Section 3. Later, in Section 4 we particularize our abstract model into two different models (multinomial and Dawid–Skene) when annotations are selected out of a discrete set of possible labels. In Section 5 we show how our mathematical model can be used in a specific application scenario to evaluate the data quality of different communities performing the annotation. In addition, we conduct predictive inference about the quality of data in hypothetical scenarios within these very same communities. Finally, Section 6 concludes and discusses future work.

2. Related Work

The problem of label aggregation in crowdsourcing has received a lot of attention from the statistics and machine learning communities, starting from classical latent class models [11]. A succinct review of latent class analysis can be found in [12]. A well-known specific application of latent class analysis to label aggregation is the seminal work of Dawid and Skene [13]. Another one is the simpler multinomial model presented in [14]. Passonneau and Carpenter [15] highlight the relevance of relying on an annotation model for the analysis of crowdsourcing data. Paun et al. [14] provide a comparison of different probabilistic annotation models for the task, and conclude that using partially pooled models, such as the hierarchical Dawid–Skene model, results in very good performance among different datasets and applications.

From the methodological perspective, perhaps a more mature alternative is the CrowdTruth framework [16,17]. The conceptual departure point for this framework is the paper by Aroyo and Welty [18], which provides a good overview of the usual misconceptions in label aggregation. Their arguments are aligned with those in this paper, in particular with respect to the myths that there is "only one truth" and that "one expert is enough". Dumitrache et al. [19] provide specifics on the quality metrics used in CrowdTruth, which incorporate the ambiguities and the inter-dependency of the crowd, input data and annotations. However, although intuitive and useful, these specifics lack a strong probabilistic and information theoretic background, which is the approach taken in this work.

Bu et al. [20] propose a graph model to handle both single- and multiple-step classification tasks and try to "infer the correct label path". They present an "adapted aggregation method" for three existing inference algorithms, namely 'majority voting', 'expectation-maximization' [21] and 'message-passing' [22].

Recently, Nguyen et al. [23] have presented the CLARA framework, which is in production at Facebook and relies on a Bayesian probabilistic model. In their work, they show that the consensus achieved by CLARA is clearly better than that obtained by majority voting. However, the application scenario is restricted to a binary labeling problem, whilst in citizen science scenarios many-valued labeling is frequently found.

Given the fact that even the subject-matter experts can disagree for a given task (e.g., the diagnosis of a patient), the quality of the data generated by crowdsourcing needs to be ensured, or at least soundly measured. Some of the obvious reasons for this need are that the participating citizens are not necessarily experts in the field, and that the protocol itself that they are asked to follow in processing the task may lead to mistakes. Accordingly, the citizen science community has put considerable effort into defining the dimensions of data quality (e.g., [24]) and building effective strategies to improve the quality (e.g., [25,26]). This is still a hot research topic and new methods are continuously being suggested.

Since the skills of each participating citizen may vary, efficient methods of task assignation have also been studied to minimize the cost and/or maximize the accuracy of a crowdsourcing project (e.g., [22,27]).

3. Modeling the Domain

In this section we provide our conceptual and mathematical model of the citizen science crowdsourcing domain. The conceptual modeling presented in Section 3.1 is deeply influenced by the CrowdTruth framework [16]. We start by using an example to help capture the three most relevant concepts. After the conceptual model has been presented, we turn it into a probabilistic model in Section 3.2.

3.1. Participating Concepts: Tasks, Workers and Annotations

We start with an example in the disaster management area to help us make concrete the intervening concepts. Timely and accurate management when a natural disaster occurs is of fundamental importance to diminish the humanitarian impact of the disaster. However, timely and accurate disaster management requires the presence of an information system reporting the places that have been more damaged and the specifics of the support required. Information is usually scarce when a disaster occurs and social networks have been shown to provide a wealth of images and videos describing details of the disaster. Hence, structuring that information is of foremost importance for adequate disaster management. Software is already available for selecting tweets with images that potentially contain valuable information (e.g., [28]). Although computer vision advances in the last decade are astonishing, it is still the case that even the best AI-based software for the task is still far from resulting in high-quality selection and classification of the images. Fortunately, citizen scientists can help in this endeavor. Each of the images obtained from the social network by the AI can be distributed to a set of citizen scientists that can help labeling that image either as irrelevant for the task at hand or with the degree of damage observed, measuring the degree of damage as either no-damage, minimal, moderate or severe. After this annotation process takes place, we need to reconcile different labels from citizen scientists for each image and use that information to report to the disaster relief organization.

The main concepts in our model are as follows.

Worker A worker is any of the participants in the annotation process. In our example, each of the volunteer citizen scientists involved in labeling images is a worker.

Task A task can be understood as the minimal piece of work that can be assigned to a worker. In our example, labeling each of the images obtained from Twitter is a task.

Annotation An annotation is the result of the processing of the task by the worker. An example of annotation in the above described disaster management example conveys the following information: *Task 22 has been labeled by worker 12 as moderate.*

3.2. Abstract Mathematical Model

Next, we present a mathematical model of the problem. We start with a finite set of w workers $W = [1..w]$, a finite set of t tasks $T = [1..t]$ and a finite set of a annotations

$A = [1..a]$. We assume that two different workers can have different features. So, to characterize each worker we introduce a set $\mathcal{W}$, which we refer to as *worker feature space*, and a feature mapping $f_W : W \to \mathcal{W}$. Thus, for each worker $w \in W$, $f_W(w)$ provides a description of the worker capabilities and characteristics. Following the same pattern, we assume that two different tasks can also each have distinctive features, and introduce a set $\mathcal{T}$ (the *task feature space*) and a mapping $f_T : T \to \mathcal{T}$ so that for each task $t \in T$, $f_T(t)$ provides a description of the task characteristics.

Each annotation is the result of the processing of a task by a worker. Function $w_A : A \to W$ maps each annotation to its worker and function $t_A : A \to T$ maps each annotation to its task. Finally, we also introduce a set $\mathcal{A}$, which we refer to as *annotation feature space* and mapping $f_A : A \to \mathcal{A}$ so that for each $a \in A$, $f_A(a)$ describes the annotation characteristics. The spaces and mappings introduced above and represented in Figure 1 provide a backbone on which we can build.

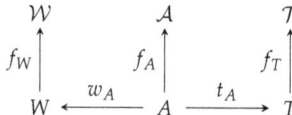

Figure 1. Spaces and mappings in our mathematical model.

The main reason for requesting the annotation of tasks by workers is that we are interested in determining a specific characteristic (or a set of characteristics) of the task that is (are) unknown to us. We assume that we can factor the task feature space $\mathcal{T}$ as $\mathcal{T} = \mathcal{T}_O \times \mathcal{T}_C \times \mathcal{T}_H$, where $\mathcal{T}_O$ contains the observable characteristics, $\mathcal{T}_C$ contains those unobservable characteristics in which we are interested and $\mathcal{T}_H$ contains those characteristics of the tasks that are unobservable and in which we are not interested in the consensus. Epistemologically, we model our lack of knowledge by means of a probability distribution, and hence we are interested in determining a probability distribution over $\mathcal{T}_C \times \overset{t}{\ldots} \times \mathcal{T}_C$. We call such a distribution a *joint consensus*. In its most general form, a joint consensus allows dependencies between the consensuses of different tasks. In this paper we will restrict our interest to individual task consensuses; that is, the marginal distribution of the joint consensus for each task.

Similarly, we can split the worker feature space $\mathcal{W}$ as $\mathcal{W} = \mathcal{W}_O \times \mathcal{W}_H$, where $\mathcal{W}_O$ contains the observable characteristics and $\mathcal{W}_H$ contains the unobservable characteristics.

We also assume the existence of some general characteristics that are relevant for the annotation. Here, the word 'general' is used in the sense of *not directly related to a specific task or a specific annotator*. We represent the domain characteristics by an element $d \in \mathcal{D}$, where we refer to $\mathcal{D}$ as the *domain space*.

3.3. The Consensus Problem

The conceptual framework introduced above allows us to properly define what a *consensus problem* is. It takes as inputs the following:

- The number of workers w, tasks t and annotations a
- For each worker $w \in W$, its observable characteristics, namely $\mathbf{w}_O^w$
- For each task $t \in T$, its observable characteristics, namely $\mathbf{t}_O^t$
- For each annotation $a \in A$, the task being annotated ($t_A(a)$), the worker that did the annotation ($w_A(a)$) and the annotated characteristics ($\mathbf{a}^a = f_A(a)$)
- A probabilistic model of annotation, consisting of the following:
  - An emission model $p(\mathbf{a}|\mathbf{w}^w, \mathbf{t}^t, \mathbf{d})$, returning the probability that in a domain of characteristics $\mathbf{d}$, a worker with characteristics $\mathbf{w}^w$ annotates a task with characteristics $\mathbf{t}^t$ with label $\mathbf{a}$.

- A joint prior over every unobservable characteristic

$$p(\mathbf{w}_H^1, \ldots, \mathbf{w}_H^w, \mathbf{d}, \mathbf{t}_C^1, \ldots, \mathbf{t}_C^t, \mathbf{t}_H^1, \ldots, \mathbf{t}_H^t).$$

Provided with the input of a consensus problem, our probability distribution factorizes as shown in Figure 2 and we can write

$$p(\mathbf{w}_H^1, \ldots, \mathbf{w}_H^w, \mathbf{d}, \mathbf{t}_C^1, \ldots, \mathbf{t}_C^t, \mathbf{t}_H^1, \ldots, \mathbf{t}_H^t, \mathbf{a}^1, \ldots, \mathbf{a}^{\mathbf{a}}) =$$

$$p(\mathbf{w}_H^1, \ldots, \mathbf{w}_H^w, \mathbf{d}, \mathbf{t}_C^1, \ldots, \mathbf{t}_C^t, \mathbf{t}_H^1, \ldots, \mathbf{t}_H^t) \prod_{a=1}^{\mathbf{a}} p(\mathbf{a}^a | \mathbf{w}^{w_A(a)}, \mathbf{t}^{t_A(a)}, \mathbf{d}).$$

It is important to highlight that, similarly to what happens in latent class models, our model encodes a conditional independence assumption. Our model assumes that annotations are independent from one another provided that we are given all the characteristics of the task, the domain and the worker. However, the conditional independence assumption encoded in latent class models is much stronger. They assume that annotations are independent provided that we know the task label. This assumption has been widely identified as a drawback of this model (as argued for example in [29], where some approaches for overcoming this drawback are also presented). By incorporating dependence on the available characteristics of the task and the annotator, the independence assumption included in our framework is much milder and justifiable. That said, the simpler incarnations of our generic model used later in the paper to analyze the Albania earthquake are in fact latent class models and, as such, encode a strong conditional independence assumption.

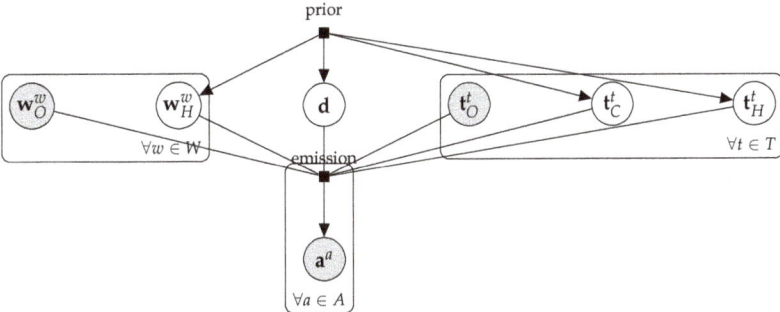

Figure 2. Probabilistic graphical model description of the abstract consensus model.

The objective of a consensus problem is answering a probabilistic query to this probability distribution. For example, finding the joint consensus $p(\mathbf{t}_C^1, \ldots, \mathbf{t}_C^t)$, which can be done by marginalizing out every hidden variable except the consensus variables $\mathbf{t}_C^1, \ldots, \mathbf{t}_C^t$.

4. Discrete Annotation Models

The abstract probabilistic model presented above is overly general and can be particularized in many different ways. Here, as in most of the cases in the literature, we assume that workers are requested to select from a finite set of annotations (the annotation feature space $\mathcal{A}$ is restricted to be a finite set). That is, when presented with a task, each annotator will annotate it by selecting an element $\mathbf{a} \in \mathcal{A} = \{\mathbf{a}_1, \ldots, \mathbf{a}_k\}$. Following the disaster management example in Section 3.1, imagine that the problem at hand is the classification of images into one of five different categories: irrelevant, no-damage, minimal, moderate or severe. In that case, we will request each worker to annotate an image by selecting a label from $\mathcal{A} = \{\text{irrelevant}, \text{no-damage}, \text{minimal}, \text{moderate}, \text{severe}\}$. Furthermore, in discrete annotation models, each task is considered to have an unobservable characteristic: its "real" label. That is, $\mathcal{T}_C = \mathcal{A}$. Next, we will see how different discrete annotation models can be accommodated into our framework.

4.1. The Multinomial Model

Perhaps the simplest model of annotation is the pooled multinomial model [14]. In the pooled multinomial model:

1. Tasks are indistinguishable other than by their real classes. That is, the task feature space $\mathcal{T} = \mathcal{T}_O \times \mathcal{T}_C \times \mathcal{T}_H$ with $\mathcal{T}_O = \mathcal{T}_H = \{\emptyset\}$ and $\mathcal{T}_C = \mathcal{A}$.
2. The general domain characteristics store the following:
   - The probability that the "real" label of a task comes from each of the classes. Thus, the domain space $\mathcal{D}$ is $\Delta\mathcal{A}$, that is, the set of probability distributions over $\mathcal{A}$. This domain can be encoded as a stochastic vector τ of dimension k, where τ_i can be understood as the probability of a task being of class a_i.
   - The noisy labeling model for this simple model is the same for each worker. It has as characteristic an unobservable stochastic matrix π of dimension $|\mathcal{A}| = k$. Intuitively, element $\pi_{i,j}$ of the matrix can be understood as "the probability that a worker labels an image of class a_i with label a_j." Thus, when π is the identity matrix, our workers are perfect reporters of the real label. The further away from the identity, the bigger the confusion.
3. Workers are indistinguishable from one another. Hence, $\mathcal{W} = \emptyset$.
4. The emission model in this case is $p(a_j|\mathbf{w},\mathbf{t},\mathbf{d}) = p(a_j|\mathbf{d} = \langle\tau,\pi\rangle, \mathbf{t}_C = a_i) = \pi_{i,j}$
5. The prior is assumed to be a Dirichlet both for τ and for each of the rows of π and also encodes that τ is the prior for the real label of the tasks.

$$p(\mathbf{w}_H, \mathbf{d}, \mathbf{t}_C) = p(\mathbf{d}) \prod_{t=1}^{t} p(\mathbf{t}_C^t | \mathbf{d}) = p(\tau) p(\pi) \prod_{t=1}^{t} p(\mathbf{t}_C^t | \tau) \qquad (1)$$

where $p(\tau) = Dirichlet(\tau; \mathbf{1}^k)$, $p(\pi) = \prod_{i=1}^{k} p(\pi_{i,\cdot}) = \prod_{i=1}^{k} Dirichlet(\pi_{i,\cdot}; \mathbf{1}^k)$, and $p(\mathbf{t}_C = a_i | \tau) = \tau_i$.

The main problem from the pooled multinomial model originates from the assumption that workers are indistinguishable from each other. This assumption is dropped in the model presented below.

4.2. The DS Model

Dawid and Skene [13] proposed one of the seminal models for crowdsourcing, the Dawid–Skene (DS) model. In this section we see how to map the DS model to our abstract framework. The DS model draws inspiration from the multinomial model, but instead of having a single noisy labeling model, it introduces a noisy labeling model per worker. Thus, in the DS model

1. Tasks are indistinguishable other than by their real classes.
2. The general domain characteristics store only the stochastic vector τ with the probability that the "real" label of a task comes from each of the classes.
3. Each worker w has as characteristic an unobservable stochastic $k \times k$ matrix π^w. Intuitively, element $\pi_{i,j}^w$ of the matrix can be understood as "the probability of worker w labeling an image of real class a_i with label a_j."
4. The emission model in this case is $p(a_j|\mathbf{w},\mathbf{t},\mathbf{d}) = p(a_j|\mathbf{w},\mathbf{t}) = p(a_j|\mathbf{w}_H = \pi^w, \mathbf{t}_C = a_i) = \pi_{i,j}^w$
5. The prior is assumed to be a Dirichlet both for p and for each of the rows of π. (The DS model, as presented in [13], used maximum likelihood to estimate its parameters and thus no prior was presented. Later, Paun et al. [14] presented the prior provided here.)

$$p(\mathbf{w}_H, \mathbf{d}, \mathbf{t}_C, \mathbf{t}_H) = p(\mathbf{w}_H) p(\mathbf{d}) p(\mathbf{t}_C|\mathbf{d}) = \prod_{w=1}^{w} p(\pi^w) p(\tau) \prod_{t=1}^{t} p(\mathbf{t}_C^t | \tau) \qquad (2)$$

where $p(\tau)$, and $p(\mathbf{t}_C|\tau)$ are the same as for the multinomial model and $p(\pi^w) = \prod_{i=1}^{k} p(\pi_{i,\cdot}^w) = \prod_{i=1}^{k} Dirichlet(\pi_{i,\cdot}^w; \mathbf{1}^k)$.

5. Evaluating Data Quality in Highly Uncertain Scenarios

In the previous section we presented a general framework and particularized it into two well-known specific models. Now, we provide an example of how our framework can be actioned to help data quality analysis in citizen science projects.

Usually, three standard methods are used for ensuring the reliability of citizen science data:

- *Gold sets* measure accuracy by comparing annotations to a ground truth;
- *Auditing* measures both accuracy and consistency by having an expert review the labels;
- *Consensus*, or overlap, measures consistency and agreement amongst a group.

We are particularly interested in building a methodology for quality assurance in scenarios, such as disaster response, where gold sets would be unavailable. In such domains, auditing may be used under the hypothesis of infallible experts (e.g., [30]). However, in Section 5.2, we show that this hypothesis does not necessarily hold in every domain. Thus, our approach is to rely on our probabilistic consensus model introduced above. We describe how to do it by following an example.

In Section 5.1, we detail the process of data collection from expert, volunteer and paid worker communities. Section 5.2 analyzes inter-expert agreement. In Section 5.3, we further scrutinize the error rates of the experts. For this evaluation, we employ our multinomial model to build their consensus and noisy labeling model. Sections 5.4 and 5.5 exploit the experts' consensus using it as the *ground truth* for the analysis of the quality of data collected from the volunteer and paid worker communities. This analysis allows us to compare the performances of these two communities, which is increasingly being discussed within the citizen science research community (e.g., [31]). Furthermore, in Section 5.6, we show how our probabilistic model can also be leveraged for a predictive analysis to estimate the number of annotations required to reach the desired accuracy for each of the three communities we worked with.

5.1. Data and Methodology

To form the set of tasks for citizen scientists, we used the social media data that were collected right after the earthquake that struck Albania on 26 November 2019. It was the strongest earthquake to hit the country in more than 40 years and the world's deadliest in 2019 (https://en.wikipedia.org/wiki/2019_Albania_earthquake, accessed on 25 February 2021).

The extraction of the disaster information from Twitter, its filtering and automatic classification were all carried out by the AIDR image processing system [28,30], which collected data during four consecutive days following the earthquake. Out of 9241 collected tweets, AIDR produced a dataset of 907 images that it deemed relevant.

Since we lacked a gold set for our domain of interest, first we needed to establish a ground truth. Hence, we contacted a group of ten experts with prior knowledge of disaster response and crisis data. We presented the set of 907 images as tasks to the experts via the Crowd4EMS platform [32]. Crowd4EMS combines automatic methods for gathering information from social media and crowdsourcing techniques, in order to manage and aggregate volunteers' contributions. We asked our experts to assess the severity of the damage on each image and annotate it with one of the five labels given in Section 4. To account for the possibility that the experts may also suffer from biases, each image was evaluated by three experts. Specifically, images were ordered and then they were presented to the experts following a single sequence, independently of the expert who was requesting. That is, the first image was assigned to the first image request received, the second image was assigned to the second request and so on until the complete set of images was assigned a label. The only constraint was that an image was not presented to the same expert more than once. This process was repeated three times, guaranteeing that each image was labeled at least three times by different experts.

Then, we provided the same dataset to a community of fifty volunteers via Crowd4EMS and to paid workers via the Amazon Mechanical Turk (MTurk) platform.

MTurk is an online micro-tasking platform that allows requesters to distribute tasks that are difficult to classify for machine intelligence, yet simple for humans, to a large group of users termed as workers for a monetary incentive. Both crowds were asked to annotate the images using the above-mentioned set of five labels. The same image assignment process detailed for experts was followed for the volunteers as well so that each image was labeled at least by three different volunteers, who made 3015 contributions in total. In MTurk, 171 paid workers participated in labeling and each image was annotated by ten different workers, thus a total of 9070 contributions was made by this community.

5.2. Evaluating Expert Infallibility

Since the annotation data collected from the experts was destined to be used as the ground truth, before any further analysis, we measured the inter-expert agreement of their answers in two ways. First, we checked the percentage of full agreement (i.e., when all annotators annotated an image with the same label). Second, we calculated the Fleiss generalized Kappa coefficient [33], which is a well-known inter-rater reliability measure. We saw that the experts fully agreed only 61.41% on the image labels. The Fleiss generalized Kappa coefficient for their answers was 53.6%. According to Landis and Koch [34], this Kappa statistic corresponds to a "moderate" agreement (for Landis and Koch [34], a Kappa statistic has to be ≥ 0.81 to be considered "almost perfect").

As a conclusion of this analysis we have to drop the expert infallibility assumption in this domain. This raises the question about how to better evaluate the data quality of a volunteer or paid crowd for a specific task, particularly when we cannot assume expert infallibility. To this end, in the next section, we propose a simple methodology that is based on the probabilistic labeling model that we introduced in Section 4.1.

5.3. Error Rates for Experts

Given the experts' annotation data, we use the multinomial model presented in Section 4.1 for three purposes: (1) to draw the joint consensus for the labeled tasks, i.e., $p(t_C^1, \ldots, t_C^t)$ where $t = 907$; (2) to estimate the $k \times k$ stochastic matrix π with $k = 5$ that characterizes the single noisy labeling model for all experts; and (3) to estimate the a priori probabilities of each label (i.e., τ). This process is illustrated in the upper half of Figure 3, where the upper box corresponds to the expectation-maximization (EM) algorithm applied to annotation data by Dawid and Skene [13]. We particularized this EM algorithm into our multinomial model. Specifically, the initial estimates of the joint consensus are calculated as raw probabilities of each label for each task using the annotation data. Then, in the Maximization step, the algorithm calculates the maximum likelihood estimates of π and τ using the current estimate of the joint consensus (with Equations (2.3) and (2.4) in [13], respectively). Subsequently, in the Expectation step, current estimates of π and τ are used to calculate the new estimates of the joint consensus (with Equation (2.5) in [13]). The algorithm alternates between these two steps until a desired convergence for the joint consensus is achieved or a given number of iterations is reached. The final estimates of π and τ are values that maximize the full likelihood of the annotation data, and the final estimates of the joint consensus gives us $p(t_C^1, \ldots, t_C^t)$.

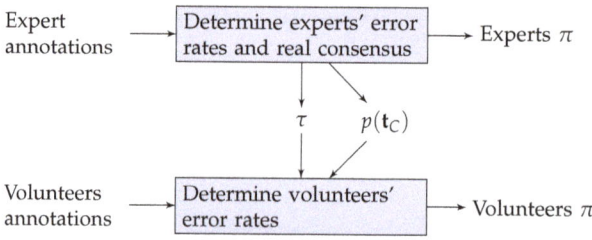

Figure 3. Flow of data for computing the error rates of the experts and volunteers.

As detailed in Section 4.1, the consensus is essentially a t × k matrix where the $(t,i)^{th}$ entry is the probability of the real label of task t (t_C^t) being a_i. Figure 4 presents the error rates of experts for whom we detailed their calculation in the above paragraph. Each cell in the figure contains the corresponding $\pi_{i,j}$ value; that is, the probability that an expert will report label a_j for an image when the real label is a_i. For experts, the real label of an image is the label with the highest probability. Although the cells include $j = i$ pairs, which are the correctly reported labels, this matrix is known as the error rate (a.k.a. confusion matrix) in the literature.

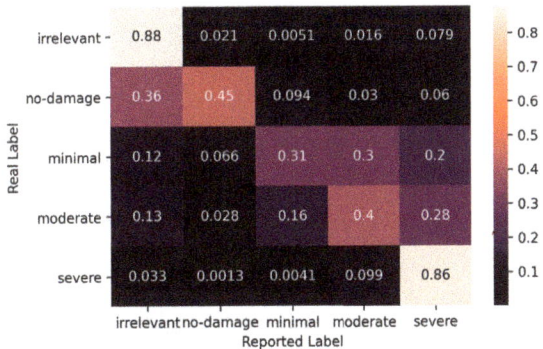

Figure 4. Error rates for experts.

In Figure 4, we observe that although irrelevant and severe damage labels are more likely to be correctly annotated, the expected accuracies for other answers are drastically low. Each of the moderate, minimal and no-damage labels has a probability below 50% of being correctly reported by the experts. For example, moderate damage is estimated to be correctly labeled only with a 40% probability, and mislabeled as severe 28% of the time. In addition, in the case of minimal damage, it is confused with moderate almost with the same probability.

As mentioned in Section 4.1, the ideal π would be an identity matrix whereas for our group of experts and set of tasks, Figure 4 is far from being an identity matrix. Thus, for our domain of interest, the experts' π also corroborates the exclusion of the expert infallibility assumption that we have shown in the above section. Experts do disagree.

In the following two sections, we will use experts' consensus as the ground truth for the evaluation of the quality of labeling made by the volunteer and paid workers. Then, following these sections we will use the a priori probability vector τ for a prospective data quality analysis for all three communities.

5.4. Evaluation of Volunteer Crowd

Given the experts' joint consensus, we fit our probabilistic model to the volunteer crowd's labeling data, and calculated their error-rates as illustrated in the lower half of Figure 3. The lower box corresponds to Equation (2.3) in [13], which calculates the maximum likelihood estimate of the π for volunteers by using the ground truth we achieved from our model for the experts in the above section. The error rates for labels are given in Figure 5.

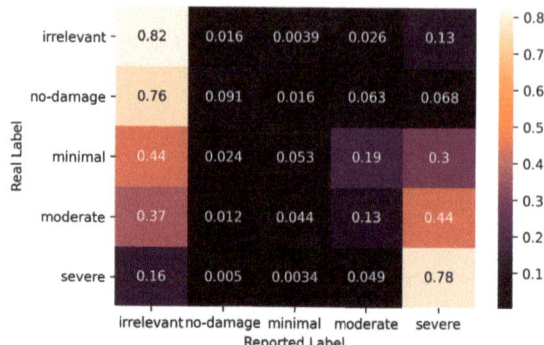

Figure 5. Error rates for volunteer crowd.

The figure shows that the volunteers have a lower probability of correct labeling compared to the corresponding probabilities for experts given in Figure 4. In particular, the probabilities of correct labels for minimal and no-damage are near zero. In both cases, the volunteers are likely to label them as irrelevant. On the other hand, moderate damage will be labeled correctly only with a 13% probability, while it is more likely to be regarded as either severe or irrelevant with probabilities of 44% and 37%, respectively.

We can speculate that the apparent subjectivity in volunteer labels may be due to the fact that each image is only annotated by 3.32 volunteers on average, and that more annotations could be expected to yield more accuracy. This speculation is exactly what we will be analyzing in Section 5.6.

5.5. Evaluation of Paid Crowd

In a similar way to the volunteers, given the experts' consensus, we fit our probabilistic model to the paid crowd's labeling data by using this data as an input to the lower box in Figure 3, which calculated, this time, the maximum likelihood estimate of the π for this crowd given the ground truth we computed via experts' annotation data in Section 5.3. The calculated error rates of paid workers are shown in Figure 6.

In the figure, we see that paid workers do a similar 'good' job like volunteers for the severe damage. They also approximate to the experts' performance for moderate and minimal damage, and no-damage. However, paid workers do seem to fail for the irrelevant images, and they are even expected to label them as severe 33% of the time.

Section 5.6 will help us to speculate on the cost-effectiveness of paid workers compared to that of the volunteers as we analyze the expected accuracies of both communities for higher number of annotations.

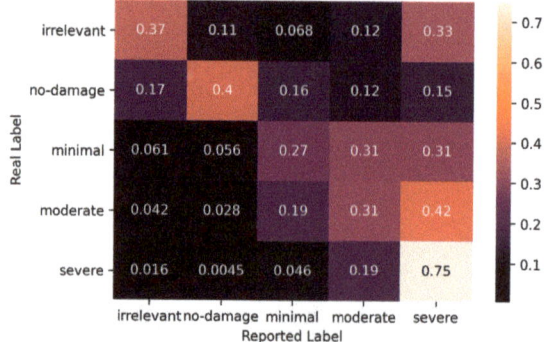

Figure 6. Error rates for paid crowd.

5.6. Prospective Comparison

In this section we aim to make a predictive analysis for the accuracy that we would expect from our three communities when they contribute with more annotations per image. For this purpose, we synthetically generated labeling-data for each of the expert, volunteer and paid communities for varying numbers of annotators by using the parameters of their probabilistic models.

Specifically, we had a separate multinomial model that we fit to each one of the data of three communities in Sections 5.3–5.5. As we know from Section 4.1, a multinomial model allows us to estimate the a priori probabilities of each label (i.e., τ), and the probability distribution of the error rates for an annotator (i.e., π) for the corresponding crowd. Accordingly, we created two synthetic sets with 1000 and 10,000 tasks. The *synthetic real labels* of the tasks were assigned by following the τ that we obtained from the model for experts in Section 5.3. The τ for experts was calculated as [irrelevant:0.3361, no-damage: 0.0271, minimal:0.0218, moderate:0.0469, severe:0.5681].

Then, for each community, by following the π of the community model, we synthetically generated a different set of labeling-data that corresponded to each of the given number of annotations. Subsequently, for each item of synthetic community data, we calculated the *synthetic consensus* of the community for the corresponding number of annotations.

Finally, by using the set of *synthetic real* labels and *synthetic consensuses*, we calculated the accuracy of all (crowd, number of annotations) pairs. The accuracy was calculated as the percentage of correct labels, and the label attached by the crowd to an image was selected as the label with the highest probability in the corresponding consensus.

Furthermore, to be able to compare the performance of the multinomial model to the standard majority voting method, we also measured the accuracy when the *synthetic consensus* was calculated by the latter method instead.

Figure 7 depicts the accuracy in the assessment of the severity of damage by the three communities for different numbers of tasks and different numbers of annotations per task. In all sub-figures, we observe that the probabilistic model outperforms majority voting for the corresponding community. The performance is comparable only for the lowest number of annotations per task—which was three—for the expert and volunteer communities in Figure 7a–d.

The figures show that when we use the proposed probabilistic model for consensus, the probability of mislabeling decreases as the number of annotators increases. This observation is also true for majority voting, but this aggregation method converges to a certain percentage after which no increase is achieved no matter how many annotations per task are carried out. Majority voting results in less accurate data, especially for the paid workers, as can be seen in Figure 7e,f. This is probably due to the poor performance of the paid crowd for the irrelevant images as we examined in Section 5.5, as these images may form an important part (expected to be 33.6%) of the synthetic real labels according to the the experts' a priori probability distribution of labels.

We also note that, although the average accuracy increases with the number of annotations in Figure 7c, we observe that the uncertainty around the accuracy also increases as opposed to other plots in Figure 7. This is due to the fact that the EM algorithm used to calculate the synthetic consensus for prospective analysis is prone to getting stuck in local maxima in the case of volunteers as a consequence of their error rates. As seen in Figure 5, the values on the diagonal are very low for three labels, and this causes the label emitted by our model to be frequently different from the real label for these classes. Hence, it is very difficult for the EM algorithm not to get stuck in alternative local maxima instead of reaching the global maximum (which ought to be close to the error-rates shown in Figure 5).

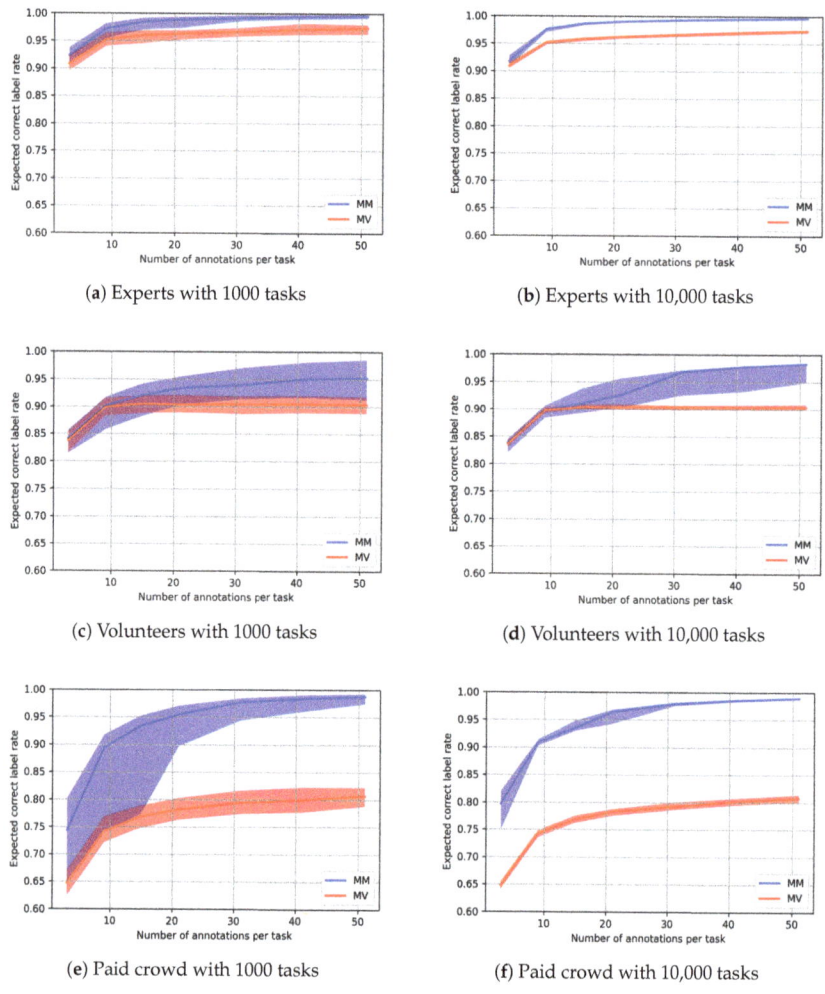

Figure 7. Prospective analysis for correct label rates for experts, volunteers and paid crowd with multinomial model (MM) and majority voting (MV).

Finally, we can say that with the probabilistic model, we do not only achieve more accurate consensus results, but we also achieve them with less annotators. Therefore, our probabilistic model is more cost-effective than the standard majority voting scheme.

6. Conclusions and Future Work

In this work we have introduced a conceptual and formal framework for modeling crowdsourced data obtained in citizen science projects. We have shown how this general model can encompass different probabilistic models already presented in the literature, such as the multinomial or the Dawid–Skene models. Finally, we have seen a use case of application to citizen science data obtained for disaster management purposes, by modeling the data obtained to perform damage assessment of the 2019 Albania earthquake. We have seen that our probabilistic model helps build a methodology that

- can be applied in scenarios where the hypothesis of infallible experts does not hold;
- can be used to characterize and study the different behaviors of different communities (in our case, experts, volunteers and paid workers);

- can be actioned to perform prospective analysis, allowing the manager of a citizen science experiment to make informed decisions on aspects such as the number of annotations required for each task to reach a specific level of accuracy.

Our work is only a first step towards establishing a scientific methodology for the analysis of crowdsourced citizen science data. In the future, we plan to fit into this very same conceptual and formal framework the application of active learning strategies (e.g., [35]) for coordinating to which workers each task should be sent to minimize the number of annotations necessary.

We have started experimenting with the use of Bayesian methods with the objective of obtaining more realistic prospective analyses. We have identified that a major problem for the application of generic inference platforms (such as Stan [36]) is the label-switching problem [37]. However, while crowdsourcing models can be understood as discrete mixture models, we think that the particularities of the task can be used to build models that are free from label switching and we are working towards proving the usefulness of those models.

In damage assessment scenarios such as the one reported in this paper, we could also model the severity of damage as a fuzzy linguistic variable with five fuzzy labels. It could be interesting to study whether crowdsourcing information could be used to learn the membership function for each fuzzy label, thus modeling uncertainty through a vague description of the concepts instead of (or in combination with) modeling the error introduced by the workers.

Author Contributions: Conceptualization, J.C., J.H.-G., A.R.S. and J.L.F.-M.; Data curation, A.R.S.; Investigation, J.C. and M.O.M.; Methodology, J.C., M.O.M. and A.R.S.; Project administration, J.L.F.-M.; Software, J.C. and M.O.M.; Supervision, J.L.F.-M.; Validation, J.H.-G.; Writing—original draft, J.C. and M.O.M.; Writing—review & editing, J.C., M.O.M., J.H.-G. and J.L.F.-M. All authors have read and agreed to the published version of the manuscript.

Funding: This work was partially supported by the projects Crowd4SDG and Humane-AI-net, which have received funding from the European Union's Horizon 2020 research and innovation program under grant agreements No 872944 and No 952026, respectively. This work was also partially supported by the project CI-SUSTAIN funded by the Spanish Ministry of Science and Innovation (PID2019-104156GB-I00). We acknowledge the support of the publication fee by the CSIC Open Access Publication Support Initiative through its Unit of Information Resources for Research (URICI). J.H.-G. is a Serra Húnter Fellow.

Institutional Review Board Statement: Not applicable.

Informed Consent Statement: Not applicable.

Data Availability Statement: We plan to make the data available through Zenodo.

Acknowledgments: We want to thank Muhammad Imran, from Qatar Computing Research Institute, for sharing their pre-filtered social media imagery dataset on the Albanian earthquake from the Artificial Intelligence for Disaster Response (AIDR) Platform. We want to thank Hafiz Budi Firmansyah, Lecturer from Sumatra Institute of Technology for his collaboration in setting up the activation on the Amazon Mechanical Turk Platform. We would also like to extend our gratitude to the volunteers for their contribution on the Crowd4EMS Platform.

Conflicts of Interest: The authors declare no conflict of interest.

References

1. Gura, T. Citizen science: Amateur experts. *Nature* **2013**, *496*, 259–261. [CrossRef]
2. Haklay, M. Citizen Science and Volunteered Geographic Information: Overview and Typology of Participation. In *Crowdsourcing Geographic Knowledge: Volunteered Geographic Information (VGI) in Theory and Practice*; Sui, D., Elwood, S., Goodchild, M., Eds.; Springer: Dordrecht, The Netherlands, 2013; pp. 105–122. [CrossRef]
3. González, D.L.; Alejandrodob; Therealmarv; Keegan, M.; Mendes, A.; Pollock, R.; Babu, N.; Fiordalisi, F.; Oliveira, N.A.; Andersson, K.; et al. Scifabric/Pybossa: v3.1.3. 2020 Available online: https://zenodo.org/record/3882334 (accessed on 16 August 2020).
4. Lau, B.P.L.; Marakkalage, S.H.; Zhou, Y.; Hassan, N.U.; Yuen, C.; Zhang, M.; Tan, U.X. A survey of data fusion in smart city applications. *Inf. Fusion* **2019**, *52*, 357–374. [CrossRef]

5. Fehri, R.; Bogaert, P.; Khlifi, S.; Vanclooster, M. Data fusion of citizen-generated smartphone discharge measurements in Tunisia. *J. Hydrol.* **2020**, *590*, 125518. [CrossRef]
6. Kosmidis, E.; Syropoulou, P.; Tekes, S.; Schneider, P.; Spyromitros-Xioufis, E.; Riga, M.; Charitidis, P.; Moumtzidou, A.; Papadopoulos, S.; Vrochidis, S.; et al. hackAIR: Towards Raising Awareness about Air Quality in Europe by Developing a Collective Online Platform. *ISPRS Int. J. Geo-Inf.* **2018**, *7*, 187.10.3390/ijgi7050187. [CrossRef]
7. Feldman, A.M. Majority Voting. In *Welfare Economics and Social Choice Theory*; Feldman, A.M., Ed.; Springer: Boston, MA, USA, 1980; pp. 161–177. [CrossRef]
8. Moss, S. *Probabilistic Knowledge*; Oxford University Press: Oxford, UK; New York, NY, USA, 2018.
9. Shannon, C.E. A mathematical theory of communication. *Bell Syst. Tech. J.* **1948**, *27*, 379–423. [CrossRef]
10. Cover, T.M.; Thomas, J.A. *Elements of Information Theory*, 2nd ed.; John Wiley & Sons: Hoboken, NJ, USA, 2006.
11. Collins, L.M.; Lanza, S.T. *Latent Class and Latent Transition Analysis: With Applications in the Social, Behavioral, and Health Sciences*; Wiley Series in Probability and Statistics; Wiley: New York, NY, USA, 2009.
12. He, J.; Fan, X. Latent class analysis. *Encycl. Personal. Individ. Differ.* **2018**, *1*, 1–4.
13. Dawid, A.P.; Skene, A.M. Maximum Likelihood Estimation of Observer Error-Rates Using the EM Algorithm. *J. R. Stat. Soc. Ser. C (Appl. Stat.)* **1979**, *28*, 20–28. [CrossRef]
14. Paun, S.; Carpenter, B.; Chamberlain, J.; Hovy, D.; Kruschwitz, U.; Poesio, M. Comparing Bayesian Models of Annotation. *Trans. Assoc. Comput. Linguist.* **2018**, *6*, 571–585. [CrossRef]
15. Passonneau, R.J.; Carpenter, B. The Benefits of a Model of Annotation. *Trans. Assoc. Comput. Linguist.* **2014**, *2*, 311–326.10.1162/tacl_a_00185. [CrossRef]
16. Inel, O.; Khamkham, K.; Cristea, T.; Dumitrache, A.; Rutjes, A.; van der Ploeg, J.; Romaszko, L.; Aroyo, L.; Sips, R.J. CrowdTruth: Machine-Human Computation Framework for Harnessing Disagreement in Gathering Annotated Data. In *The Semantic Web—ISWC 2014*; Mika, P., Tudorache, T., Bernstein, A., Welty, C., Knoblock, C., Vrandečić, D., Groth, P., Noy, N., Janowicz, K., Goble, C., Eds.; ; Lecture Notes in Computer Science; Springer: Cham, Switzerland, 2014; pp. 486–504. [CrossRef]
17. Dumitrache, A.; Inel, O.; Timmermans, B.; Ortiz, C.; Sips, R.J.; Aroyo, L.; Welty, C. Empirical methodology for crowdsourcing ground truth. *Semant. Web* **2020**, 1–19. [CrossRef]
18. Aroyo, L.; Welty, C. Truth Is a Lie: Crowd Truth and the Seven Myths of Human Annotation. *AI Mag.* **2015**, *36*, 15–24. [CrossRef]
19. Dumitrache, A.; Inel, O.; Aroyo, L.; Timmermans, B.; Welty, C. CrowdTruth 2.0: Quality Metrics for Crowdsourcing with Disagreement. *arXiv* **2018**, arXiv:1808.06080.
20. Bu, Q.; Simperl, E.; Chapman, A.; Maddalena, E. Quality assessment in crowdsourced classification tasks. *Int. J. Crowd Sci.* **2019**, *3*, 222–248. [CrossRef]
21. Dempster, A.P.; Laird, N.M.; Rubin, D.B. Maximum Likelihood from Incomplete Data via the EM Algorithm. *J. R. Stat. Soc. Ser. B (Methodol.)* **1977**, *39*, 1–38.
22. Karger, D.R.; Oh, S.; Shah, D. Iterative Learning for Reliable Crowdsourcing Systems. In *Advances in Neural Information Processing Systems 24*; Shawe-Taylor, J., Zemel, R.S., Bartlett, P.L., Pereira, F., Weinberger, K.Q., Eds.; Curran Associates, Inc.: New York, NY, USA, 2011; pp. 1953–1961.
23. Nguyen, V.A.; Shi, P.; Ramakrishnan, J.; Weinsberg, U.; Lin, H.C.; Metz, S.; Chandra, N.; Jing, J.; Kalimeris, D. CLARA: Confidence of Labels and Raters. In Proceedings of the 26th ACM SIGKDD International Conference on Knowledge Discovery & Data Mining, KDD '20, New York, NY, USA, 23–27 August 2020; Association for Computing Machinery: New York, NY, USA, 2020; pp. 2542–2552. [CrossRef]
24. Pipino, L.L.; Lee, Y.W.; Wang, R.Y. Data Quality Assessment. *Commun. ACM* **2002**, *45*, 211–218. [CrossRef]
25. Freitag, A.; Meyer, R.; Whiteman, L. Strategies Employed by Citizen Science Programs to Increase the Credibility of Their Data. *Citiz. Sci. Theory Pract.* **2016**, *1*, 2. [CrossRef]
26. Wiggins, A.; Newman, G.; Stevenson, R.D.; Crowston, K. Mechanisms for Data Quality and Validation in Citizen Science. In Proceedings of the IEEE Seventh International Conference on e-Science Workshops, Stockholm, Sweden, 5–8 December 2011; pp. 14–19. [CrossRef]
27. Ho, C.J.; Vaughan, J. Online Task Assignment in Crowdsourcing Markets. In Proceedings of the AAAI Conference on Artificial Intelligence, Toronto, ON, Canada, 22–26 July 2012; Volume 26.
28. Imran, M.; Castillo, C.; Lucas, J.; Meier, P.; Vieweg, S. AIDR: Artificial intelligence for disaster response. In Proceedings of the 23rd International Conference on World Wide Web, Seoul, Korea, 7–11 April 2014; pp. 159–162.
29. van Smeden, M.; Naaktgeboren, C.A.; Reitsma, J.B.; Moons, K.G.M.; de Groot, J.A.H. Latent Class Models in Diagnostic Studies When There is No Reference Standard—A Systematic Review. *Am. J. Epidemiol.* **2014**, *179*, 423–431. [CrossRef]
30. Imran, M.; Alam, F.; Qazi, U.; Peterson, S.; Ofli, F. Rapid Damage Assessment Using Social Media Images by Combining Human and Machine Intelligence. *arXiv* **2020**, arXiv:2004.06675.
31. Kirilenko, A.P.; Desell, T.; Kim, H.; Stepchenkova, S. Crowdsourcing analysis of Twitter data on climate change: Paid workers vs. volunteers. *Sustainability* **2017**, *9*, 2019. [CrossRef]
32. Ravi Shankar, A.; Fernandez-Marquez, J.L.; Pernici, B.; Scalia, G.; Mondardini, M.R.; Di Marzo Serugendo, G. Crowd4Ems: A crowdsourcing platform for gathering and geolocating social media content in disaster response. *Int. Arch. Photogramm. Remote Sens. Spat. Inf. Sci.* **2019**, *42*, 331–340. [CrossRef]

33. Gwet, K.L. *Handbook of Inter-Rater Reliability: The Definitive Guide to Measuring the Extent of Agreement Among Raters*, 4th ed.; Advanced Analytics, LLC: Gaithersburg, MD, USA, 2014.
34. Landis, J.R.; Koch, G.G. The Measurement of Observer Agreement for Categorical Data. *Biometrics* **1977**, *33*, 159–174. [CrossRef] [PubMed]
35. Sheng, V.S.; Provost, F.; Ipeirotis, P.G. Get another label? improving data quality and data mining using multiple, noisy labelers. In Proceeding of the 14th ACM SIGKDD International Conference on Knowledge Discovery and Data Mining—KDD 08, Las Vegas, NV, USA, 24–27 August 2008; ACM Press: New York, NY, USA, 2008; p. 614. [CrossRef]
36. Carpenter, B.; Gelman, A.; Hoffman, M.D.; Lee, D.; Goodrich, B.; Betancourt, M.; Brubaker, M.; Guo, J.; Li, P.; Riddell, A. Stan: A Probabilistic Programming Language. *J. Stat. Softw.* **2017**, *76*, 1–32. [CrossRef]
37. Rodríguez, C.E.; Walker, S.G. Label Switching in Bayesian Mixture Models: Deterministic Relabeling Strategies. *J. Comput. Graph. Stat.* **2014**, *23*, 25–45. [CrossRef]

Article

Visualizing Profiles of Large Datasets of Weighted and Mixed Data

Aurea Grané * and Alpha A. Sow-Barry

Statistics Department, Universidad Carlos III de Madrid, 28903 Getafe, Spain; alpha.jp96@yahoo.com
* Correspondence: aurea.grane@uc3m.es

Abstract: This work provides a procedure with which to construct and visualize profiles, i.e., groups of individuals with similar characteristics, for weighted and mixed data by combining two classical multivariate techniques, multidimensional scaling (MDS) and the *k*-prototypes clustering algorithm. The well-known drawback of classical MDS in large datasets is circumvented by selecting a small random sample of the dataset, whose individuals are clustered by means of an adapted version of the *k*-prototypes algorithm and mapped via classical MDS. Gower's interpolation formula is used to project remaining individuals onto the previous configuration. In all the process, Gower's distance is used to measure the proximity between individuals. The methodology is illustrated on a real dataset, obtained from the Survey of Health, Ageing and Retirement in Europe (SHARE), which was carried out in 19 countries and represents over 124 million aged individuals in Europe. The performance of the method was evaluated through a simulation study, whose results point out that the new proposal solves the high computational cost of the classical MDS with low error.

Keywords: clustering; Gower's interpolation formula; Gower's metric; mixed data; multidimensional scaling

1. Introduction

One of the most important goals in visualizing data is to get a sense of how near or far objects are from each other. Often, this is done with a scatter plot, because the Euclidean distance is the only one that our brain can easily interpret. However, scatter plots cannot always be obtained from raw data, nor is the Euclidean distance always the appropriate one to be computed on raw data. This may be the case when comparing a high number of variables, where a dimension reduction is usually necessary to better see the proximities between objects, or when working with more complex datasets, such as weighted mixed data or functional data, where other distances are preferred to the Euclidean one. For instance, survey data coming from macro-surveys at national and cross-national levels are rather complex datasets of weighted and mixed data. They are composed of variables of different natures, such as binary, multi-state categorical and numerical variables; and as result of a multi-stage sampling methodology, they each include a weighting variable, so that each individual represents a group of different size for the target population. Another added complexity may be their large or very large sample size (10^4 or larger).

Multidimensional scaling (MDS) is one of the most extended methodologies to analyze and visualize the profile structure of data, and can address some of those problems. This dimensionality reduction technique takes a dissimilarity or distance matrix as the input and produces a pictorial representation of the data in a Euclidean space, similar to a scatter plot. An important limitation when working with large and very large datasets is that it relies on the eigendecomposition of the full distance matrix between objects or individuals, thereby requiring large quantities of memory and very long computing times.

Paradis (2018) [1] proposed an approach to avoid the limitations of the standard MDS procedure, which is based on a random selection of a small number of observations and the application of standard MDS with one or two dimensions. In a second step, the remaining observations were projected and several algorithms were proposed and studied. Some drawbacks were pointed out in the discussion of the paper, such that procedures were tested on 100 points chosen randomly, since a larger value would make them slower and more complicated, and one of the approaches does not seem a viable solution to handle datasets larger than 10^4.

The main objective of this work is to provide a procedure to construct and visualize profiles, i.e., groups of individuals with similar characteristics, for weighted and mixed data by combining two classical multivariate techniques, MDS and the k-prototypes clustering algorithm [2]. Since classical MDS suffers from computational problems as sample size increases, we propose instead a "fast" MDS based on the selection of a small random sample, which is clustered by means of an adapted version of the k-prototypes algorithm that can cope with Gower's metric and weighted data. At the same time, the selected sample is mapped onto an MDS configuration and the remaining objects are projected onto the previous configuration via Gower's interpolation formula. The profile visualization is achieved by assigning each projected object to the closest cluster's centroid and coloring it accordingly. Finally, profile main characteristics are computed as the "average" member of each cluster, where the mode is considered for categorical variables and the means or the medians for quantitative ones. We give a flowchart with an overview of the algorithm steps in Figure 1.

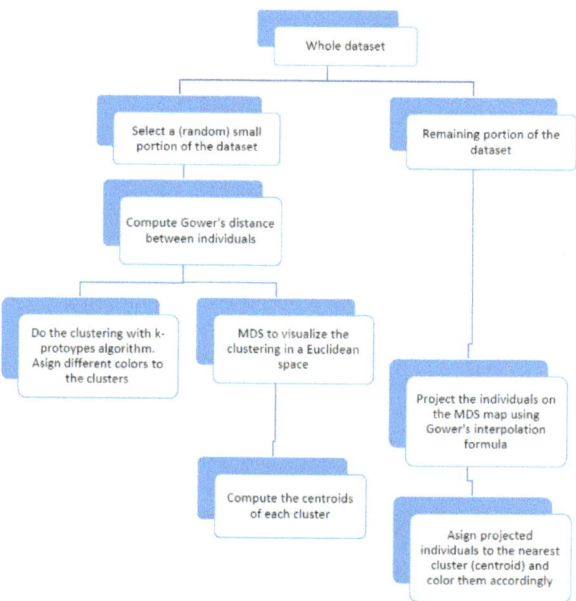

Figure 1. Flowchart of the visualization algorithm.

Note that our proposal starts by clustering the individuals in the original space, and next, we use MDS to visualize the clustering in the Euclidean space. For that reason we use a clustering algorithm able to cope with mixed data, k-prototypes (although other methods can be used [3,4]). Another possibility would be to start with the MDS representation and next do the clustering on the Euclidean space using k-means clustering. In any case, when working with large datasets, a "fast" MDS is required in order to reduce both computational time and memory use. Thus, in any case, the idea of applying the methodology to a small portion of the data and to project the remaining individuals on the MDS-map remains. In this work, we explore the former; that is, we cluster the individuals in the original space instead of doing the clustering on the MDS-map, since an additional motivation was to incorporate Gower's distance in the k-prototypes clustering algorithm.

The performance of our proposal was evaluated through a simulation study based on a real dataset of weighted and mixed data that came from the Survey of Health, Ageing and Retirement in Europe (SHARE), carried out in 19 countries. The analysis was applied to 60,020 individuals, who represent a target population of more than 124 million Europeans aged 55 years or over.

The work proceeds as follows: In Section 2 we review weighted MDS; in Section 3 we present the proposed methodology; in Section 4.1 we apply the method to data coming from SHARE database; Section 4.2 contains the simulation studies; and we conclude in Section 5.

2. Materials and Methods

In this section we give a general overview of classical MDS for weighted mixed data, introduce some useful notation and present Gower's interpolation formula.

The purpose of MDS is to construct a set of points in a Euclidean space whose interdistances are either equal (classical MDS) or approximately equal (ratio, interval or ordinal MDS) to those in a given matrix of dissimilarities, so that the interpoint distances approximate the interobject dissimilarities. That is, given an $n \times n$ matrix $\mathbf{D}^{(2)} = (\delta_{ij}^2)_{1 \leq i,j \leq n}$, where δ_{ij}^2 is the squared dissimilarity between objects i, j, for $i, j = 1, \ldots, n$, the objective of MDS is to search for a configuration of n points on a set of orthogonal axes, so that the l^2-distances between the coordinates of these n points coincide with the corresponding entries in $\mathbf{D}^{(2)}$. These coordinates are called a Euclidean configuration/map/representation or MDS configuration/map/representation of $\mathbf{D}^{(2)}$. Several possible measures of approximation between interpoint distances and interobject dissimilarities can be used, each yielding a different MDS configuration. In this work, these coordinates are obtained via spectral decomposition. General contextual references are [5–8].

Sample surveys often employ multistage sampling schemes which involve unequal selection probabilities at some or all stages of the sampling process. We refer to this situation as a weighted context, where each individual can represent a population group of different size. In this framework, classical methods of data analysis which assume simple random sampling may no longer be valid, and weighting may appear as the only or best alternative. Albarrán et al. (2015) [9] reviewed the extension of classical MDS concepts to the weighted context.

2.1. Weighted MDS

Let $\{\mathbf{x}_i, i = 1, \ldots, n\}$ be n p-dimensional vectors which contain the observations or measurements of p variables for n different individuals and $\mathbf{D}^{(2)}$ be the matrix of squared distances between n individuals, with entries $\delta^2(\mathbf{x}_i, \mathbf{x}_j)$, $1 \leq i,j \leq n$. Remember that the information contained in the p variables can be either of a quantitative or qualitative nature, or both, hence it is crucial to select an appropriate dissimilarity function in the computation of $\mathbf{D}^{(2)}$ in order to incorporate all the statistical information contained in the data.

Additionally, since each individual in the dataset can represent a group of a different size of the target population (weighted context), we have a vector of weights $\mathbf{w} = (w_1, \ldots, w_n)'$, such that $w_i > 0$, for $i = 1, \ldots, n$, and $\mathbf{1}'\mathbf{w} = 1$, where $\mathbf{1}$ is the $n \times 1$ vector of 1 s. Note that in the case of simple random sampling—that is, when each individual in the dataset represents a group of the same size of the target population—$w_i = 1/n$, for $i = 1, \ldots, n$, and classical MDS formulae are recovered.

Suppose that we are interested in obtaining an MDS representation of $\mathbf{D}^{(2)}$, provided that $\mathbf{D}^{(2)}$ satisfies the Euclidean requirement.

Given $\mathbf{w}$, we define an $n \times n$ diagonal matrix $\mathbf{D_w} = \text{diag}(\mathbf{w})$, whose diagonal is the vector of weights, and $\mathbf{J_w} = \mathbf{I} - \mathbf{1}\mathbf{w}'$ is the w-centering matrix, where $\mathbf{I}$ is the $n \times n$ identity matrix.

The w-centering matrix $\mathbf{J_w}$ is an orthogonal projector with respect to $\mathbf{D_w}$, idempotent (that is, $\mathbf{J_w^2} = \mathbf{J_w}$) and self-adjoint with respect to $\mathbf{D_w}$ (that is, $\mathbf{J_w'D_w} = \mathbf{D_wJ_w}$).

In the weighted context, the doubly w-centered inner product matrix is given by

$$\mathbf{G_w} = -\frac{1}{2}\mathbf{J_w D}^{(2)} \mathbf{J_w'},$$

whose standardized version is given by

$$\mathbf{F_w} = \mathbf{D_w^{1/2} G_w D_w^{1/2}}, \quad (1)$$

which is called standardized inner product matrix. The condition that $\mathbf{D}^{(2)}$ satisfies the Euclidean requirement is equivalent to imposing that $\mathbf{G_w}$ is positive semidefinite, which means that there exists a matrix $\mathbf{Y_w}$ such that $\mathbf{G_w} = \mathbf{Y_w Y_w'}$. In the weighted context, $\mathbf{Y_w}$ is called a w-centered Euclidean representation of $\mathbf{D}^{(2)}$ and satisfies the following two properties:

(a) $\mathbf{w'Y_w} = \mathbf{0}$.
(b) The squared l^2-distances between the rows of $\mathbf{Y_w}$ coincide with the corresponding entries in $\mathbf{D}^{(2)}$; that is, for each pair of individuals i, j, we have that $(\mathbf{y_{w,i}} - \mathbf{y_{w,j}})'(\mathbf{y_{w,i}} - \mathbf{y_{w,j}}) = \delta^2(\mathbf{x_i}, \mathbf{x_j})$ where $\mathbf{y_{w,i}}$ is the i-th row of $\mathbf{Y_w}$.

Matrix $\mathbf{Y_w}$ is the w-weighted MDS representation of $\mathbf{D}^{(2)}$ and is computed by means of the spectral decomposition of matrix $\mathbf{F_w}$ defined in Equation (1). That is, given $\mathbf{F_w} = \mathbf{U\Lambda U'}$, where $\mathbf{\Lambda}$ is a diagonal matrix with the eigenvalues of $\mathbf{F_w}$, ordered in descending order, and $\mathbf{U}$ is the corresponding matrix of eigenvectors (in column),

$$\mathbf{Y_w} = \mathbf{D_w^{-1/2} U \Lambda^{1/2}}, \quad (2)$$

whose rows are the principal coordinates of n individuals, and its columns are the principal axes of this representation.

2.2. Gower's Distance

Gower's similarity coefficient [10] is one of the most popular similarity measures and perhaps the easiest way to obtain a distance measure when working with mixed data. It is the Pitagorean sum of three similarity coefficients, one for each type of variable. In particular, it uses Jaccard's coefficient for binary variables, the simple matching coefficient for multi-state categorical variables and range-normalized city block distance for quantitative variables. Given two p-dimensional vectors $\mathbf{x_i}$ and $\mathbf{x_j}$, Gower's similarity coefficient is defined as:

$$s(\mathbf{x_i}, \mathbf{x_j}) = \frac{\sum_{h=1}^{p_1}\left(1 - |x_{ih} - x_{jh}|/R_h\right) + a + \alpha}{p_1 + (p_2 - d) + p_3},$$

where p_1 is the number of quantitative variables, R_h is the range of the h-th quantitative variable, a is the number of positive matches, d is the number of negative matches for the p_2 binary variables, α is the number of matches for the p_3 multi-state categorical variables and $p = p_1 + p_2 + p_3$. The entries of matrix $\mathbf{D}^{(2)}$ are computed as:

$$\delta^2(\mathbf{x_i}, \mathbf{x_j}) = 1 - s(\mathbf{x_i}, \mathbf{x_j}), \quad (3)$$

and Gower (1971) [10] proved that Equation (3) satisfies the Euclidean requirement.

Other metrics for mixed data can be considered, although the k-prototypes algorithm should be modified accordingly. For instance, a more robust metric that can overcome some of the shortcomings of Gower's is related metric scaling (RelMS) by Cuadras (1998) [11], which was used in [9] to obtain robust profiles in weighted and mixed datasets. However, in this work, we prefer to illustrate our methodology by using Gower's metric due to the computational complexity of RelMS.

2.3. Gower's Interpolation Formula

A very useful tool to project new data points onto a given MDS configuration is Gower's interpolation formula [8], which was extended to the weighted context by [12]. The following proposition can be derived from the Theorem 1 in [12]. This formula is a key tool in the visualization algorithm that we propose.

Proposition 1. Let $\mathcal{E}$ be a set of n individuals; $\mathbf{w} = (w_1, \ldots, w_n)'$ a vector of weights, such that $w_i > 0$, for $i = 1, \ldots, n$, and $\mathbf{1}'\mathbf{w} = 1$; and $\mathbf{D}^{(2)}$ be a matrix of squared distances between the n individuals satisfying the Euclidean requirement and $\mathbf{Y_w}$ the $\mathbf{w}$-weighted metric scaling representation of $\mathbf{D}^{(2)}$.

Given a new individual $n+1$ for whom the squared distances to n individuals of $\mathcal{E}$ are known, $\delta = (\delta_{n+1,1}^2, \ldots, \delta_{n+1,n}^2)$, its principal coordinates can be computed as:

$$\mathbf{y}_{n+1} = \frac{1}{2}(\mathbf{g_w} - \delta)\mathbf{D_w Y_w \Lambda}^{-1}, \quad (4)$$

where $\mathbf{g_w} = \text{diag}(\mathbf{G_w})'$ is a row vector containing the diagonal elements of $\mathbf{G_w}$, $\mathbf{D_w} = \text{diag}(\mathbf{w})$, $\mathbf{G_w} = \mathbf{Y_w Y_w'}$ and Λ is the diagonal matrix containing the eigenvalues of matrix $\mathbf{F_w}$ defined in (1).

Proof. The squared distance between the individual $n+1$ and any individual $i \in \mathcal{E}$ is given by

$$\delta_{n+1,i}^2 = (\mathbf{y}_{n+1} - \mathbf{y}_i)(\mathbf{y}_{n+1} - \mathbf{y}_i)' = \mathbf{y}_{n+1}\mathbf{y}_{n+1}' - 2\mathbf{y}_{n+1}\mathbf{y}_i' + \mathbf{y}_i\mathbf{y}_i'.$$

In matrix notation, we have that

$$\delta = \|\mathbf{y}_{n+1}\|^2 \mathbf{1}' - 2\mathbf{y}_{n+1}\mathbf{Y_w'} + \mathbf{g_w}. \quad (5)$$

Post multiplying expression (5) by $\mathbf{D_w Y_w}$ and after operating, we have that:

$$2\mathbf{y}_{n+1}\mathbf{Y_w' D_w Y_w} = (\mathbf{g_w} - \delta)\mathbf{D_w Y_w} + \|\mathbf{y}_{n+1}\|^2 \mathbf{1}' \mathbf{D_w Y_w}.$$

Note that $\|\mathbf{y}_{n+1}\|^2 \mathbf{1}' \mathbf{D_w Y_w} = 0$ since $\mathbf{1}' \mathbf{D_w} = \mathbf{w}'$ and $\mathbf{w}' \mathbf{Y_w} = 0$. Therefore, the principal coordinates of individual $n+1$ are given by:

$$\mathbf{y}_{n+1} = \frac{1}{2}(\mathbf{g_w} - \delta)\mathbf{D_w Y_w}(\mathbf{Y_w' D_w Y_w})^{-1} = \frac{1}{2}(\mathbf{g_w} - \delta)\mathbf{D_w Y_w \Lambda}^{-1},$$

since from Formula (2) we have that $\mathbf{Y_w' D_w Y_w} = \Lambda^{1/2}\mathbf{U}'\mathbf{D_w}^{-1/2}\mathbf{D_w D_w}^{-1/2}\mathbf{U}\Lambda^{1/2} = \Lambda$. □

3. Methodology

In this section we discuss a methodology for visualizing profiles for large datasets of weighted and mixed data.

Among all the approaches proposed for visualizing data, MDS is one of the most common techniques. However, we find the classical MDS algorithm a limited tool when visualizing large datasets, since it requires very large CPU time or large computing memory when dealing with large distance matrices. Delicado and Pachón-García (2020) [13] showed both: that the time needed to compute MDS as a function of the sample size increases notably when using *cmdscale()* R function (in stats package by R Development Core Team) and that at least 400 MB of RAM memory is required to store the distance matrix when there are 10,000 observations.

There have been several attempts to solve the scalability problem, such as steerable multidimensional scaling [14], incremental MDS [15], relative MDS [16], FastMap [17], MetricMap [18], landmark MDS [19], the diagonal majorization algorithm [20] and uniform manifold approximation and projection [21]. SteerMDS proposed by Williams et al. [14] is based on a spring-mass model, introduced by Chalmers [22] (see also [23] for a sampling-based variant of the algorithm). These methods calculate lower-dimensional coordinates by iteratively minimizing a cost or stress function that is proportional to the distance between the current coordinates and the given dissimilarities. Incremental MDS is similar to the previous methods, although it focuses on overall shape instead of local details. Relative MDS combines MDS with the learning vector quantization clustering method. FastMap, MetricMap and Landmark MDS approximate classical MDS by solving MDS for a subset of the data and fit the remainder to the solution. Platt [24] studied these methods and concluded that Landmark MDS was the fastest and most accurate of the them. The diagonal majorization algorithm is a modification of the Guttman majorization algorithm [25] that is used to minimize the stress function, and is able to save computing time taking into account several factors (see [26]). Finally, the uniform manifold approximation and projection is a learning technique for dimension reduction that can be used for visualization, which is indirectly related to MDS and more closely to Isomap.

The visualization method that we propose is based on classical MDS applied to a random portion of the dataset plus the projection of the remaining individuals via Gower's interpolation formula. Thus, our proposal shares the idea behind several of the existing methods of applying MDS to a portion of the dataset, but differs from them in the projection tool used to obtain the final MDS representation.

3.1. The Visualization Algorithm

We start by summarizing the proposed method, and later we remark on some important aspects concerning the clustering algorithm and the feasible implementation of Gower's interpolation formula.

The starting point of the algorithm is a large dataset of weighted and mixed data; that is, our dataset was composed of several variables of different natures (binary, multi-state categorical and numerical variables) plus a weighting variable containing the individual weights. Remember that weights are given exogenously and are related to the survey sampling technique. No pre-processing of the data is needed, except for the normalization of weights to sum to 1.

1. Select a small random sample, using the weights to produce a more informative sample, that is, trying to follow as much as possible the sampling scheme. Depending on the size of dataset, this selection can be 2.5, 5 or 10% of total observations. Let us denote this small sample by $\mathbf{X}_{n \times p}$, where n is the number of individuals and p the number of variables.
2. Compute the distance matrix between the rows of $\mathbf{X}_{n \times p}$ using Gower's distance Formula (3).
3. Carry out the k-prototypes clustering algorithm in order to find the different clusters and label the individuals accordingly. Determine the number of clusters in the dataset by the "elbow" rule.
4. Obtain the principal coordinates of the labeled individuals through weighted MDS.
5. Compute the representatives (or centroids) of the clusters. This can be done by calculating the weighted mean or weighted median of those point-coordinates belonging to the same cluster in the MDS configuration.
6. Project the rest of the individuals (the remaining 97.5, 95 or 90%) onto the MDS configuration using Gower's interpolation formula.
7. Finally, from the MDS configuration, assign the new points to an existing cluster based on the closest centroid (according to Euclidean distance) and label/color them accordingly.

Once all points have been assigned to a cluster, it is possible to visualize the clusters on the MDS configuration, and thus, to see the proximities between them. Finally, a profile is defined as the "average" member of each cluster. To do so, for categorical variables the mode can be computed and the mean and the median are good options for quantitative variables.

3.2. Some Important Remarks

Next, we introduce a few observations on the steps presented above.

3.2.1. On the Clustering Algorithm

Classical hierarchical clustering can handle also mixed data by using Gower's coefficient. However, it is well known that it struggle as sample size increases. Reference [2] proposed a clustering algorithm called k-prototypes which is quite efficient, $O(T+1)kn$, where n is the number of observations, k the number of clusters and T the number of iterations; and it has been previously used by [27] for profile construction. The k-prototypes algorithm is based on a dissimilarity measure that takes into account both quantitative and categorical variables. Let $X_1^r, \ldots, X_{p_1}^r, X_{(p_1+1)}^c, \ldots, X_p^c$ be the variables available in the dataset, where X_h^r stands for a quantitative variable and X_h^c for a categorical one; then the dissimilarity between two individuals $x_i, x_j \in \mathbb{R}^p$ can be measured by

$$d_2(\mathbf{x}_i, \mathbf{x}_j) = \sum_{h=1}^{p_1} (x_{ih} - x_{jh})^2 + \gamma \sum_{h=p_1+1}^{p} \delta(x_{ih} - x_{jh}), \tag{6}$$

where $\delta(p,q) = 0$ for $p = q$ and $\delta(p,q) = 1$ for $p \neq q$, and $\gamma \geq 0$ is a coefficient that measures the influences of numeric and categorical variables. Note that when $\gamma = 0$, clustering only depends on numeric variables.

As it happens in any non-hierarchical clustering method, the number of clusters, k, must be determined in advance. To do so, a variety of techniques exist, and sometimes determining the optimal number of clusters is an inherently subjective measure that depends on the goal of the analysis. Due to the large size of the dataset, we decided to use the "elbow" method, instead of the average silhouette width or other time-consuming criteria. To apply the "elbow" method, we ran the algorithm for different values of k and calculated the cost function for each run. Then, we plotted the cost function in a line graph; and the point where a turning point (or "elbow") was observed, that is, the point at which the cost function levels off, was selected as the optimal k value. Recently, Aschenbruck and Szepannek (2020) [28] examined the transferability of cluster validation indices to mixed data and evaluated them through simulation studies. They concluded that the average silhouette width was the most suitable with respect to both runtime and determination of the correct number of clusters. However, these conclusion rely on rather small datasets (≤ 400 individuals).

In this work, we introduce two particularities to the k-prototypes algorithm so that it can cope with Gower's distance and weighted datasets.

First, instead of the d_2 measure described in (6), we used Gower's distance Formula (3), which is a very popular dissimilarity measure for mixed data and satisfies the Euclidean requirement [10]. The second particularity introduced refers to weighted datasets. Since the ultimate difference from the standard algorithm is in centroid calculation, weighted averages of quantitative variables and weighted modes of categorical variables are used, instead of standard means and modes.

In what follows, we summarize the adapted version of the k-prototypes algorithm:

- Initial prototypes selection. Select k distinct individuals from the dataset as the initial centroids.
- Initial allocation. Each individual of the dataset is assigned to the closest prototype's cluster, according to distance (3).
- Reallocation. The prototypes for the previous and current clusters of the individuals must be updated, taking into account individual weights. This repeats until there is no reallocation of individuals.

Some authors pointed out possible inaccurate clustering results when using Hamming distance and proposed other alternatives ([29,30]). In our simulations, we did not experiment with such situations (see Section 4.2 for graphical representations of the cost function).

However, we propose to adapt the *k*-prototypes algorithm, although its convergence properties in combination with using Gower's distance were not further investigated.

3.2.2. On Gower's Interpolation Formula

As mentioned before, MDS's final configuration is obtained by projecting the remaining observations via Gower's interpolation formula on the initial configuration. But how can this be implemented in a feasible way? The idea is the following:

We call $\mathbf{M}_{m \times p}$ the remaining observations after selecting $\mathbf{X}_{n \times p}$.

- Split $\mathbf{M}_{m \times p}$ row-wise into ℓ partitions $\mathbf{M}_1, \ldots, \mathbf{M}_\ell$, equally sized, with perhaps the exception of $\mathbf{M}_\ell$, which can be smaller. The number of partitions is set to be $(n+m)/\ell$, where $\ell \times \ell$ is the size of the largest distance matrix that a computer can calculate efficiently [13].
- Apply Gower's interpolation formula to each matrix $\mathbf{M}_j$ ($j = 1, \ldots, \ell$) and store the coordinates. The application of Gower's interpolation formula to a matrix $\mathbf{M}_j$ whose rows are *m* "new" individuals is rather straightforward from Formula (4). With the same notation as in Section 2.3, let $\boldsymbol{\Delta}$ be the $m \times n$ matrix whose rows contain the squared distances of the "new" *m* individuals to the *n* individuals of $\mathcal{E}$. Then, the principal coordinates of these "new" *m* individuals can be computed by:

$$\mathbf{Y}_m = \frac{1}{2}(\mathbf{1}_m\, \mathbf{g_w} - \boldsymbol{\Delta})\mathbf{D_w}\mathbf{Y_w}\boldsymbol{\Lambda}^{-1},$$

where $\mathbf{1}_m$ is a $m \times 1$ vector of ones.

This is simple, but strongly advantageous, since one of the main problems in MDS is memory consumption when computing distance matrices. Gower's interpolation formula allows us to iteratively get the MDS configurations without facing memory problems, because we are reducing the size of the corresponding ℓ matrices that contain the squared distances to the *n* individuals of the existing MDS configuration. This aspect is studied in Section 4.2.

3.3. R Functions

There are several ways to perform metric MDS with R (see the MASS package for non-metric methods via the *isoMDS* function). In the following, we list them with their corresponding packages within parentheses:

- *cmdscale* (stats by R Development Core Team),
- *pcoa* (ape by [31]),
- *dudi.pco* (ade4 by [32]),
- *smacofSym* (smacof by [33]),
- *wcmdscale* (vegan by [34]),
- *pco* (labdsv by [35]),
- *pco* (ecodist by [36]).

All the functions listed above require a distance matrix as the main argument to work with. In case data are not in the distance/dissimilarity matrix format, R-functions dist, daisy and gower.dist may be of help. Moreover, some of the previous packages provide their own functions for calculating distances.

With respect to MDS configurations, the wcmdscale() function is used in this work. It is based on function cmdscale (base package of R—stats) and it can use point weights. Points with high weights will have stronger influences on the result than those with low weights. Setting equal weights will give ordinary multidimensional scaling.

Concerning the *k*-prototypes algorithm, the R package clustMixType by [37] contains the function kproto() needed to perform this technique. As mentioned before, we modified this function to achieve our desired goal, that is, to cope with Gower's distance and weighted datasets.

4. Results

In this section we present a real data application and a simulation study to evaluate the performance of the visualization algorithm.

4.1. Application

Prior to analyzing the performance of the proposed methodology, we applied the algorithm described in Section 3 to a dataset from the Survey of Health, Ageing and Retirement in Europe (SHARE). SHARE is a research infrastructure, formed by a panel database of micro data, for studying the effects of health, social, economic and environmental policies over the lifetimes of European citizens and beyond. It was founded in 2002 and is coordinated centrally at the Munich Center for the Economics of Aging (MEA), the Max Planck Institute for Social Law and Social Policy (Munich, Germany). The SHARE interview is ex-ante harmonized and all aspects of the data generation process, from sampling to translation, and from fieldwork to data processing, have been conducted according to strict quality standards. As a result, SHARE has the advantage of encompassing cross-national variations in public health and socioeconomic living conditions of European individuals, and becoming a major pillar of the European research area, with over 9000 researchers registered as SHARE users. See their website http://www.share-project.org/ for further details. This work uses Wave 6 of SHARE, which was conducted in 2015 in 18 European countries and Israel. It asked questions ranging from an individual's financial situation to his/her self-perception of health.

4.1.1. Description of the Dataset

The dataset to be analyzed consists of 60,020 observations and 13 variables, and includes a weighting variable that scales to represent over 124 million elderly individuals in Europe. Descriptions of variables are in Table 1. It is important to remark that the last four correspond to health and wellbeing indexes and were not in the original dataset, but created by [27] from the aggregations of 30 variables. Higher values of the indices correspond to situations of greater vulnerability. Although the process of creating those indices and other details about the dataset are described in their work, here we give a brief summary of the indices for better interpretation of the profiles.

The dependency index summarizes the loss of personal autonomy. It includes variables that reflect difficulties in performing activities of daily living, instrumental activities of daily living, mobility limitations and so on.

The self-perception of health index is quite a subjective measure that captures some aspects of subjective wellbeing. This index includes variables related to what an individual thinks about their health rather than their physical reality, including how satisfied the interviewee is with their life, whether they are feeling depression, etc.

The physical health and nutrition index captures the risk of an individual to suffering or developing serious health problems. It contains information related to body mass index, grip strength, nutrition and chronic conditions of the individual.

The mental agility index captures the mental acuteness of the respondents and is related to cognitive functions. It contains the results of tests around numeracy, orientation and linguistic fluency.

Next, we proceed with the visualization and construction of the profiles for this dataset of weighted and mixed data.

Table 1. Descriptive variables included in the analysis and their possible values or categories.

Type	Description	Values/Categories
CT	Country	19 countries
B	Gender	"Male", "Female"
CT	Ages	"55–60", "61–65", "66–75", "76+"
B	Employment status	"Employed", "Not working"
B	Marital status	"Has no spouse", "Has a spouse"
CT	Education	"No education", "Primary", "Secondary", "University"
B	Household in financial distress	"Yes", "No"
CT	Household receives benefits or has payments?	"Payments and no benefits", "No benefits and no payments", "Benefits and payments", "Payments and no benefits"
C	Dependency index	Form 0 to 10
C	Physical health and nutrition index	From 0 to 10
C	Self-perception of health index	From 0 to 10
C	Mental agility index	From 0 to 10

B = binary, CT = categorical, C = continuous.

4.1.2. Visualization of Profiles and Findings

The result of applying the visualization algorithm is shown in Figure 2. In order to select the number of clusters, the algorithm was run for $k = 2, \ldots, 10$. The cost function showed an "elbow" at around $k = 3$ or $k = 4$. We investigated having 3 or 4 profiles and selected $k = 4$, since $k = 3$ led to overly broad results. Panel (a) contains the MDS configuration based on Gower's interpolation formula computed from a portion of 2.5% of the data. Red triangles correspond to the mapped points of the random selection, and gray crosses stand for the projected ones. In panel (b) we show the pictorial representation of the clusters, where blue triangles stand for cluster centroids and circles represent their confidence regions, whose radii were computed as the 90th percentile of the (Euclidean) distance between each point and the corresponding centroid. We can observe beforehand two distinct groups of individuals, that is, a set of points grouped on the left (cluster 3) and another bunch of crowded points on the right, which splits into three small clusters. As will be seen later, cluster 3 corresponds to the least disadvantaged profile, whereas clusters 2 and 4 contain the most vulnerable individuals, according to the descriptive variables.

When original variables (and not just a matrix of distances or similarities) are available, it may be of interest to determine the influences of these original variables on the MDS dimensions, i.e., to determine which variables explain the most the homogeneity within groups. It was therefore necessary to calculate some correlation coefficient (or association measure) between the principal coordinates and the variables. We used Pearson's correlation coefficient for continuous variables, Spearman's correlation coefficient for ordinal variables and Cramer's V measure of association for nominal variables. Results are shown in Table 2, where it can be seen that categorical variables, such as gender, job (Employment status), fdistress (household in financial distress) and paybene (household receives benefits or has payments) have great influences on the axes. For instance, the first principal coordinate is mostly determined by the variables job (employment status) and paybene (household receives benefits or has payments), whereas fdistress (household in financial distress) and gender are influential to the second and third principal coordinates. Quantitative variables do not seem to have many impacts on the axes, although the variable mental (mental agility index) influences the second coordinate somewhat. This was expected somehow, since Gower's metric tends to give more weight to qualitative variables.

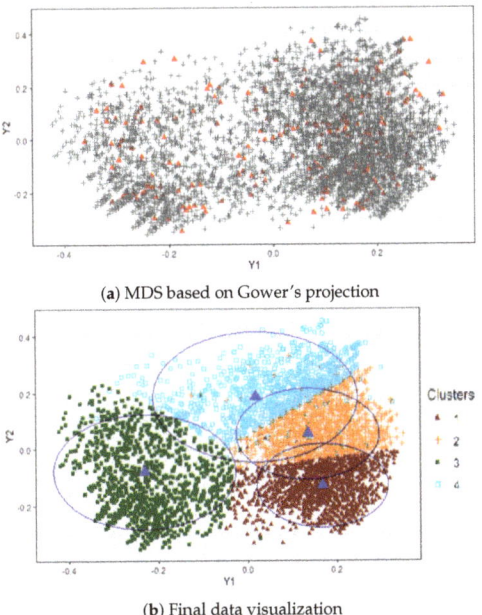

(a) MDS based on Gower's projection

(b) Final data visualization

Figure 2. Visualization of profiles. MDS based on Gower's projection and final data visualization.

Table 2. Correlations/associations among the original variables and the first three principal axes.

Variable	1st PC (Y1)	2nd PC (Y2)	3rd PC (Y3)
age	0.4192	0.1829	0.1017
gender	0.4258	0.4336	0.6147
job	0.8589	0.3522	0.1669
fdistress	0.1196	0.4597	0.7053
mstat	0.4244	0.4100	0.4248
edu	−0.0796	−0.3448	−0.2684
paybene	0.5372	0.1670	0.1004
depend	0.3577	0.3326	0.2047
health	0.3406	0.3003	0.1693
sph	0.2651	0.1473	0.2357
mental	0.1856	0.4129	0.3549

A graphical way to see the influences of the original variables on the construction of the profiles is to assign colors to the categories of the original variables (or groups of values, in the case of the quantitative ones) and represent the individuals colored accordingly in the MDS configuration. In Figure 3 we show the most representative projections of the MDS configuration regarding those variables more correlated/associated with the principal axes.

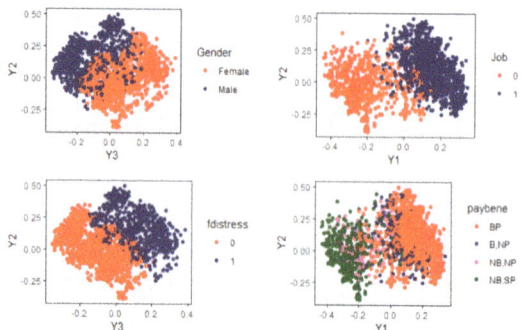

Figure 3. MDS representation. Influential variables.

In the following we summarize the main characteristics of the four estimated profiles shown in Figure 2 (right panel). The results were acquired by taking the weighted mode for qualitative variables and the weighted mean and median for quantitative ones. See also Figures 4 and 5 and Table 3.

P1: This included 30.48% of respondents, representing more than 37.89 million people; 56.56% of them were female, equally likely to belong to any age bracket, although more than 50% were under 71 years old; 55.18% were secondary-school-educated and more than 75% had secondary or university study behind them; not working; lived with a partner; health-related benefits and payments; wellbeing index mean values around 2–2.26 and median values of 2.

P2: This included 22.76% of respondents, representing more than 28.29 million people; 50.26% of them were female; more than 50% of them were 70 years old or older; 47.80% were primary-school-educated and more than 75% were primary or secondary-school-educated; not working; lived with a partner; health-related benefits and payments; wellbeing index mean values around 3–4 and median values 2–4.

P3: This included 26.82% of respondents, representing more than 33.34 million people; 57.27% female; 59.95% were between 55–65 years old and more than 80% were under 66 years old; around 70% were secondary-school-educated or university-educated; working; lived with a partner; no health-related benefits, some payments; least vulnerable group, wellbeing index mean values around 1.4–2.20, median values 0–2.

P4: This included 19.94% of respondents, representing more than 24.78 million people; 51.35% female; 40.82% were 76 years old or older and more than 50% were older than 70 years old; low education (more than 75% were primary-school-educated or not at all); not working; likely lived alone or with a partner; health-related benefits and payments; 96.69% in financial distress; most vulnerable group, wellbeing index mean values around 3.4–5.4 and median values 4–6.

From the previous description we can sort the profiles from the most to the least disadvantaged, in terms of levels of health and socioeconomic wellbeing. In particular, P4 defines the group with the lowest levels of health and wellbeing, closely followed by P2. On the other hand, P1 defines a medium–low social vulnerability profile and P3 a profile of low risk of social vulnerability.

From Figure 4 and Table 3 we observe that profiles P4 and P2 more resemble each other compared to both other profiles across "dependency," "mental agility" and "physical health," whereas "self-perception of health" is more balanced. Remember that higher values of the indices correspond to situations of greater vulnerability.

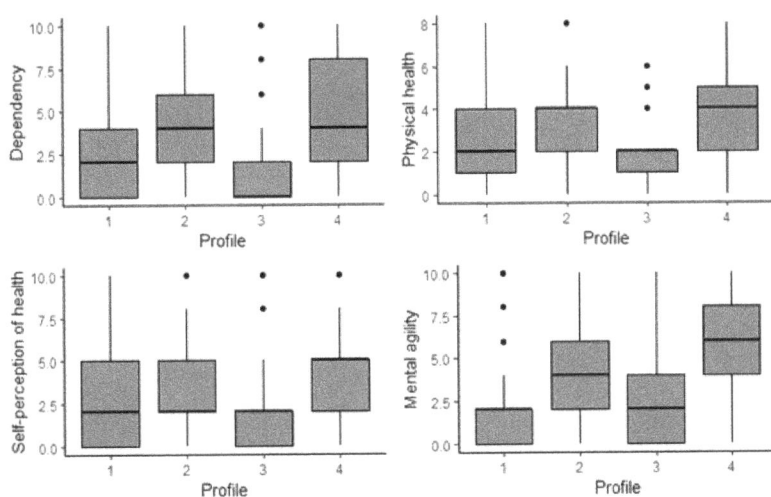

Figure 4. A boxplot distribution of the indices by profile.

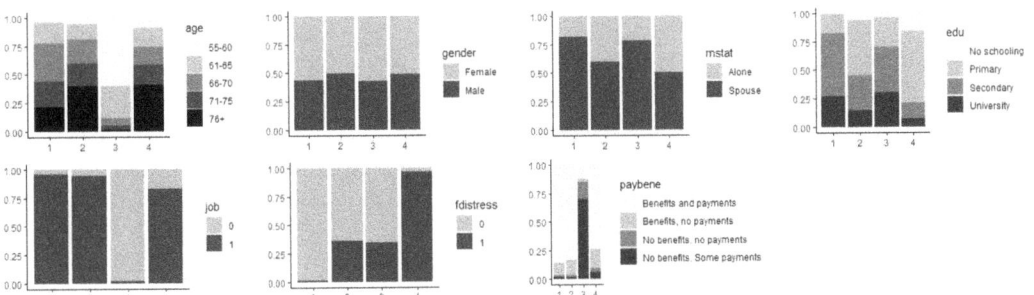

Figure 5. Distribution of descriptive variables by profile.

P2 and P4 are the most vulnerable, although P2 is still better than P4. These profiles are especially vulnerable, since they have high percentages of individuals belonging to the 76 years or older group and they rank mostly bad in the indices, mainly due to the loss of physical or mental/intellectual autonomy produced by gradual aging. However, P4 is the one that ranked worst in all wellbeing indices, with mean differences of 0.31–1.39 points with respect to P2. Moreover, people in P4 were more likely to live alone than people in P2 (49.95% in front of 39.81%). Therefore, people in P4 were of the type that require more assistance and or extensive help in order to carry out common everyday actions. Additionally, we see that 96.69% of households in P4 were in financial distress, in front of 63.88% of the households in P2. Still comparing P2 and P4, another interesting finding is that the P4 individuals were more pessimistic with respect to their health. At least, this is what the distribution of the "self-perception of health" index seems to indicate (with a median difference of three points). This variable expresses what individuals think about their health: whether they are satisfied with their life, feeling depression, etc.

Table 3. Summary statistics of the variables used to create the profiles, per profile.

Cluster	Count	% of Total	Age	Age Prop.	Gender	Gender Prop.	Job Status	Job Status Prop.
1	37,890,792.42	30.48%	66–70	33.53%	Female	56.56%	Not working	96.19%
2	28,293,780.69	22.76%	76+	39.81%	Female	50.26%	Not working	94.90%
3	33,340,913.81	26.82%	55–60	59.95%	Female	57.27%	Working	97.98%
4	24,788,136.51	19.94%	76+	40.82%	Female	51.35%	Not working	83.24%

Cluster	Financial distress	Financial distress Prop.	Marital status	Marital status Prop.	Education level	Education level Prop.	Payments or benefits?	Payments or benefits? Prop.
1	No	98.68%	Spouse	81.82%	Secondary	55.18%	B & P	85%
2	No	63.88%	Spouse	60.19%	Primary	47.80%	B & P	83%
3	No	65.17%	Spouse	78.22%	Secondary	39.44%	P & No B	70%
4	Yes	96.69%	Spouse	50.05%	Primary	63.28%	B & P	75%

Cluster	Dependency index mean	Dependency index median	Physical health index mean	Physical health index median	Self-perceived health index mean	Self-perceived health index median	Mental agility index mean	Mental agility index median
1	2.03	2	2.24	2	2.26	2	1.92	2
2	3.74	4	3.09	4	3.26	2	4.05	4
3	1.38	0	1.82	2	1.81	2	2.20	2
4	4.60	4	3.40	4	3.94	5	5.44	6

B & P = both benefits and payments; P & No B = payments and no benefits.

Clearly, the least disadvantaged group was P3, which heavily skewed towards younger individuals, since almost 60% of the individuals were aged between 55 and 60, and those over 76 made up less than 5% of this profile. In addition, 57.27% of them were women, and they tend to work until an advanced age. In addition, the indices indicate good health for this group, which may be related to the fact that continuing working in later life has been proved to be correlated with positive health outcomes [38]. Low values in wellbeing indices are consistent with the fact that people in this profile have no health-related benefits.

Finally, we see that education is one of the factors that most influences the profiles' differences. Notice that most disadvantaged profiles have in common lower education levels (secondary, primary or none), whereas the most advantaged always have a high percentage of members who attended university. This finding supports the idea that society's disparities are explained by differences in education among individuals. Education is often used as a proxy for socioeconomic status and its impact on later-life health outcomes is well researched [39,40].

4.1.3. Profiles across Europe

The variable "country," which represents the country of residence of the individuals, was not included in the process described so far. However, it seemed interesting to see how the profiles are distributed across Europe, particularly for the most disadvantaged ones. Thus, for each country we computed the percentage of people belonging to each profile. Results are shown in Figure 6, where it can be seen that a combination of the least disadvantaged profiles is predominant in most European countries, except for Portugal, Estonia and Poland.

A more interesting question is to find out whether SDG-3 of United Nations 2030 Agenda, that is, to ensure healthy lives and promote wellbeing for all at all ages, is fulfilled in these developed countries. To do so, we analyzed the percentages of the most disadvantaged profiles in these EU countries and found that they are concentrated in the Southern, Central and Eastern European countries. In particular, in countries such as Portugal, Spain, Italy and Poland it is estimated that over 20% of their population of 55 years old or older belong to P4. Regarding P2, it is estimated to be over 20% in 66.6% of the EU countries, geographically distributed in Central–Eastern European countries.

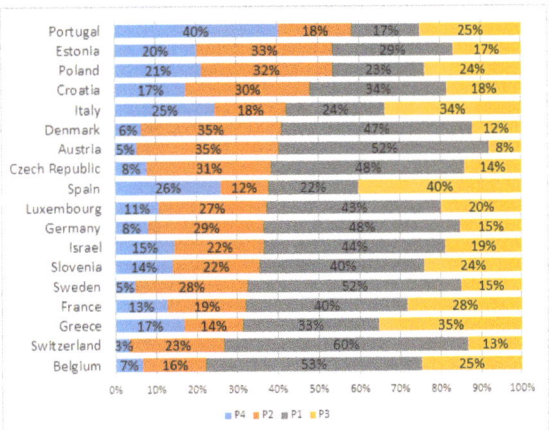

Figure 6. Distribution of P1–P4 profiles by country.

4.2. Simulation Study

Recall from Section 3 that we introduced Gower's interpolation approach as a way to solve the scalability problem. However, is this true? Have we dealt successfully with the problem? We consider those questions next. The aim of this section is to evaluate the discrepancies between two MDS configurations, the classical one computed from the complete dataset and the one obtained with our algorithm. Besides, we analyze the time required to compute both configurations.

In this simulation study, the starting point is the dataset; hence, the computing times reported in this work are not comparable with those reported in [1], where the starting points were the first two principal coordinates.

4.2.1. Design of the Simulation Study

The analysis was carried out on samples of the dataset used in Section 4.1, and it was structured as follows:

- Sample sizes. To evaluate the elapsed time, a total of nine different sample sizes were used: n = 500, 1000, 5000, 10,000, 20,000, 30,000, 40,000, 50,000 and 60,000. Discrepancies between two MDS configurations were evaluated through a total of six different sample sizes: n = 500, 1000, 2000, 3000, 4000 and 5000.
- Portion of data. Recall that the first step of the algorithm was to select a small sample from the data. We wished to see whether there exists a significant difference in using 2.5, 5 or 10% as the initial portion.
- Each scenario was the combination of a sample size and a portion of data and was repeated 100 times.

For each repetition, we computed elapsed time; the errors in configuration eigenvalues, measured in terms of mean squared error (MSE); and the cophenetic correlation between Euclidean configurations, as a measure of distortion between two distance matrices [41].

We used a personal computer to run the simulation analysis, whose technical specifications were: computer processing unit: Intel®; core™ i5-4200U CPU @ 1.60 GHz 2.30 GHz; RAM: 4 GB (Santa Clara, CA, USA).

Tables 4 and 5 contain the mean values per scenario.

4.2.2. Time to Compute MDS

As stated before, in this section we analyze the divergence of the time required to compute MDS configurations when interpolating data points and when using all points (complete MDS from now on) to construct the coordinates. Table 4 contains the simulation study results. Clearly, we see that MDS based on Gower's interpolation was, by far, the fastest one. Note that the table does not display any result for the complete MDS approach from 10^4 observations onward. This was because of memory limitation problems. It could not even store the distance matrix. Nevertheless, we see that for small sample sizes we obtained a greater elapsed time for the complete MDS case, although we were not considering the time required to get the distance matrix (if we considered this, the difference would become even larger). As sample size increases, so do the time and memory needed. For example, for sample sizes of 500 to 1000, the elapsed time was almost six times higher for the complete MDS than for Gower's approach. Another observation is that as sample size increases, it is more interesting to consider a portion of 2.5 or 5% rather than 10%. We will see in the next section whether there is a significant difference in choosing 2.5% instead of 10%.

Table 4. Time required to compute weighted MDS configurations (in seconds).

	n	Sample Portion	Gower's Interpolation	Complete MDS		n	Sample Portion	Gower's Interpolation	Complete MDS
1	500	2.5	0.16	0.45	15	20,000	10	71.89	-
2	500	5	0.19	0.55	16	30,000	2.5	30.41	-
3	500	10	0.2	0.57	17	30,000	5	91.74	-
4	1000	2.5	0.18	2.83	18	30,000	10	190.2	-
5	1000	5	0.21	2.9	19	40,000	2.5	71.4	-
6	1000	10	0.25	2.95	20	40,000	5	165.6	-
7	5000	2.5	0.87	259.52	21	40,000	10	517.2	-
8	5000	5	1.71	271.79	22	50,000	2.5	152.52	-
9	5000	10	3.01	265.14	23	50,000	5	271.5	-
10	10,000	2.5	3.91	-	24	50,000	10	1125.51	-
11	10,000	5	9.17	-	25	60,000	2.5	195.31	-
12	10,000	10	19.33	-	26	60,000	5	453.78	-
13	20,000	2.5	19.32	-	27	60,000	10	1927.23	-
14	20,000	5	29.91	-					

Table 5. Discrepancies through eigenvalues (MSE) and Euclidean configurations (cophenetic correlation).

	n	Sample Portion	Eigenvalues	Cophenetic Correlation
1	500	2.5	0.079	0.274
2	500	5	0.063	0.749
3	500	10	0.043	0.797
4	1000	2.5	0.052	0.751
5	1000	5	0.053	0.794
6	1000	10	0.037	0.825
7	2000	2.5	0.047	0.792
8	2000	5	0.037	0.821
9	2000	10	0.029	0.844
10	3000	2.5	0.038	0.811
11	3000	5	0.031	0.836
12	3000	10	0.027	0.851
13	4000	2.5	0.030	0.821
14	4000	5	0.030	0.855
15	4000	10	0.022	0.855
16	5000	2.5	0.025	0.831
17	5000	5	0.025	0.848
18	5000	10	0.020	0.858

4.2.3. Discrepancies in MDS Configurations

Here we compare the distortion of both MDS configurations by calculating the cophenetic correlation between them. That is, we want to know how different the configuration obtained through Gower's interpolation formula is from the classical MDS (complete MDS). We also analyze how much the eigenvalues of both approaches differ by means of the mean square error (MSE), computed on the normalized positive eigenvalues of the two MDS configurations. As stated before, due to memory limitations, we could not do the comparisons for sample sizes greater than 10^4 observations. Results are shown in Table 5, where we can see, first, that the normalized eigenvalues do not differ significantly (average MSE of 0.045 and median of 0.034), and second, that inter-distances between both configurations are preserved (average cophenetic correlation of 0.789 and median of 0.824). Overall, this simulation study reveals that discrepancies between configurations decrease as the sample size increases, and that, for a given sample size, they tend to decrease when we consider a higher sample portion. Although one can think of selecting a greater sample portion to reduce them, we must keep in mind that this requires more time to get the final MDS configuration.

4.2.4. Cost Function

Finally, we illustrate the convergence of the cost function of the k-prototypes algorithm, for $k = 4$ and several of the scenarios described above. In Figure 7 we depict the mean and median values of the corresponding cost functions for which convergence was achieved in a rather small number of iterations.

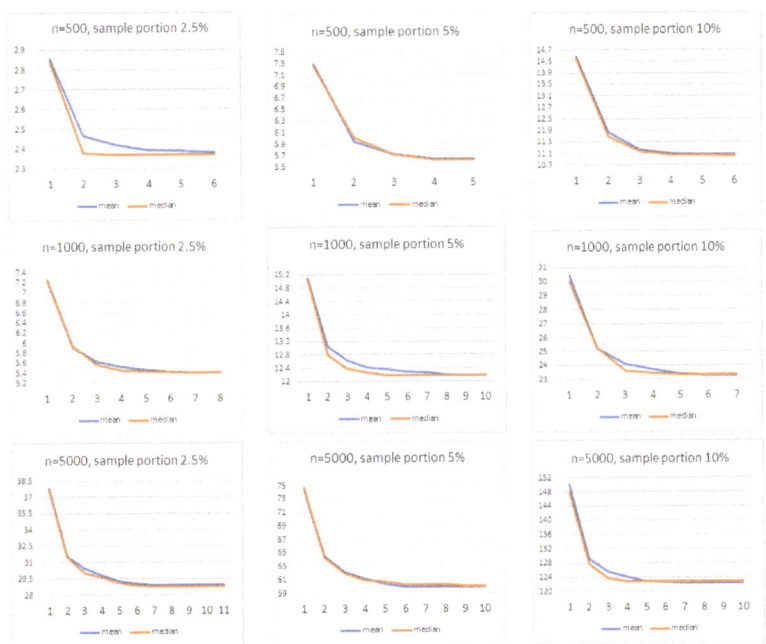

Figure 7. Cost function mean and median values.

5. Conclusions

In this paper we presented a methodology for visualizing profiles of large datasets of weighted and mixed data that combines two classical multivariate techniques, MDS and a clustering algorithm. The clustering algortihm is used to partition the individuals in the original space, and once individuals are labeled accordingly, MDS is used to visualize the clusters in a Euclidean space, where it is easier to explore the proximities among clusters. The scalability problem is solved by means of Gower's interpolation formula. Finally, profiles are obtained as the "average" member of each cluster, where the mode is considered for categorical variables and the mean or the median for numerical ones.

In this work we have illustrated this methodology by means of the k-prototypes clustering algorithm, although other clustering procedures able to cope with mixed data can be used [4]. Additional motivations for using k-prototypes were to adapt it to the weighted context and to use Gower's metric as the dissimilarity measure. However, some authors pointed out possible inaccurate clustering results when using Hamming distance and proposed other alternatives ([29,30]). However, we did not experiment with such situations in our case study.

We tested the procedure through a simulation study, where we evaluated computational costs (elapsed time, errors in configuration eigenvalues) and discrepancies between classical and Gower's interpolated MDS configurations (cophenetic correlation). The results show that MDS based on Gower's interpolation formula solves the main issue of classical MDS (high computational cost) with few errors.

We applied the proposed methodology to find several profiles of healthy life and wellbeing in the European Union, using the respondents to the Survey of Health, Ageing and Retirement in Europe survey data. We found that the most vulnerable group (people with the poorest health and wellbeing) contained elderly people who were less educated, living alone, had financial problems, had low personal autonomy and had impairments in cognitive abilities. This profile is more likely to be found in Southern and Eastern European countries.

Although Gower's metric is widely used when dealing with mixed datasets, it presents several shortcomings. For example, it gives more weight to categorical variables than to quantitative ones; it does not take into account the possible associations or correlations between variables; and it is not robust against atypical data (see [9,42,43]).

An interesting direction for future research would be to consider other metrics, such as the hybrid dissimilarity coefficient by Jian and Song [30], and their modification of the k-prototypes algorithm, although it is more time consuming than the classical k-prototypes. Nevertheless, this hybrid dissimilarity still ignores the association between variables. Another possibility would be to use related metric scaling (RelMS) to tailor the metric and incorporate it into the clustering algorithm. RelMS was introduced in [11,44] and provides more robust and stable configurations than Gower's, but it has a high computational cost, which should be reduced so that it can be workable for large datasets. Another interesting direction for future research would be to explore the possibility of combining our methodology with clustering algorithms based on archetype and archetypoid analysis, which focuses on extreme individuals instead of centroids ([45,46]).

Author Contributions: Conceptualization, A.G. and A.A.S.-B.; methodology, A.G.; software, A.A.S.-B.; validation, A.A.S.-B. and A.G., data curation, A.A.S.-B.; writing—original draft preparation, A.A.S.-B.; writing—review and editing, A.G.; supervision, A.G.; project administration, A.G.; funding acquisition, A.G. All authors have read and agreed to the published version of the manuscript.

Funding: This research was funded by the Spanish Ministry of Economy and Competitiveness, grant number MTM2014-56535-R; and the V Regional Plan for Scientific Research and Technological Innovation 2016–2020 of the Community of Madrid, an agreement with Universidad Carlos III de Madrid in the action of "Excellence for University Professors".

Institutional Review Board Statement: Not applicable.

Informed Consent Statement: Not applicable.

Data Availability Statement: Original data can be downloaded from http://www.share-project.org/.

Acknowledgments: The authors are grateful to Pedro Delicado for his useful comments and suggestions.

Conflicts of Interest: The authors declare no conflict of interest.

References

1. Paradis, E. Multdimensional scaling with very large datasets. *J. Comput. Graph. Stat.* **2018**, *27*, 935–939. [CrossRef]
2. Huang, Z. Clustering large data sets with mixed numeric and categorical values. In *Proceedings of the First Pacific Asia Knowledge Discovery and Data Mining Conference, Singapore, 23–24 February 1997*; World Scientific: Singapore, 1997; pp. 21–34.
3. Van de Velden, M.; Iodice D'Enza, A.; Markos, A. Distance-based clustering of mixed data. *Wires Comput. Stat.* **2018**, *11*, e1456. [CrossRef]
4. Ahmad, A.; Khan, S.S. Survey of State-of-the-Art Mixed Data Clustering Algorithms. *IEEE Access* **2019**, *7*, 31883–31902. [CrossRef]
5. Borg, I.; Groenen, P.J.F. *Modern Multidimensional Scaling: Theory and Applications*, 2nd ed.; Springer: New York, NY, USA, 2005.
6. Cox, T.F.; Cox, M.A.A. *Multidimensional Scaling*, 2nd ed.; Chapman and Hall: Boca Raton, FL, USA, 2000.
7. Krzanowski, W.J.; Marriott, F.H.C. *Multivariate Analysis, Part 1, Volume Distributions, Ordination and Inference*; Arnold: London, UK, 1994.
8. Gower, J.C.; Hand, D. *Biplots*; Chapman and Hall: London, UK, 1996.
9. Albarrán, A.; Alonso, P.; Grané, A. Profile identification via weighted related metric scaling: An application to dependent Spanish children. *J. R. Stat. Soc. Ser. Stat. Soc.* **2015**, *178*, 1–26. [CrossRef]
10. Gower, J.C. A general coefficient of similarity and some of its properties. *Biometrics* **1971**, *27*, 857–874. [CrossRef]
11. Cuadras, C.M. Multidimensional Dependencies in Ordination and Classification. In *Analyses Multidimensionelles des Données*; Fernández, K., Morineau, A., Eds.; CISIA-CERESTA: Saint-Mandé, France, 1998; pp. 15–25.
12. Boj, E.; Delicado, P.; Fortiana, J. Distance-based local linear regression for functional predictors. *Comput. Stat. Data Anal.* **2010**, *54*, 429–437. [CrossRef]
13. Delicado, P.; Pachón-García, C. Multidimensional Scaling for Big Data. 2020. Available online: https://arxiv.org/abs/2007.11919 (accessed on 23 July 2020).
14. Williams, M.; Munzner, T. Steerable, progressive multidimensional scaling. In *Proceedings of the Information Visualization, INFOVIS 2004, IEEE Symposium, Austin, TX, USA, 10–12 October 2004*; pp. 57–64.
15. Basalaj, W. Incremental multidimensional scaling method for database visualization. In *Proceedings of the SPIE 3643, Visual Data Exploration and Analysis VI, San Jose, CA, USA, 25 March 1999*. [CrossRef]
16. Naud, A.; Duch, W. Interactive data exploration using MDS mapping. In *Proceedings of the Fifth Conference: Neural Networks and Soft Computing, Zakopane, Poland, 6–10 June 2000*; pp. 255–260.
17. Faloutsos, C.; Lin, K. FastMap: A fast algorithm for indexing, data-mining, and visualization. In *Proceedings of the ACM SIGMOD, San Jose, CA, USA, 23–25 May 1995*; pp. 163–174.
18. Wang, J.T.-L.; Wang, X.; Lin, K.-I.; Shasa, D.; Shapiro, B.A.; Zhang, K. Evaluating a class of distance-mapping algorithms for data mining and clustering. In *Proceedings of the ACM KDD, San Diego, CA, USA, 15–18 August 1999*; pp. 307–311.
19. De Silva, V.; Tenenbaum, J.B. Global versus local methods for nonlinear dimensionality reduction. *Adv. Neural Inf. Process. Syst.* **2003**, *15*, 721–728.
20. Trosset, W.M.; Groenen, P.J. Multidimensional scaling algorithms for large data sets interactive data exploration using MDS mapping. In *Proceedings of the Computing Science and Statistics, Kunming, China, 7–9 December 2005*.
21. McInnes, L.; Healy, J.; Saul, N.; Großberger, L. UMAP: Uniform Manifold Approximation and Projection. *J. Open Source Softw.* **2018**, *3*, 861. [CrossRef]
22. Chalmers, M. A linear iteration time layout algorithm for visualizing high dimensional data. *Proc. IEEE Vis.* **1996**, 127–132. [CrossRef]
23. Morrison, A.; Ross, G.; Chalmers, M. Fast Multidimensional Scaling through Sampling, Springs, and Interpolation. *Inf. Vis.* **2003**, *2*, 68–77. [CrossRef]
24. Platt, J.C. FastMap, MetricMap, and Landmark MDS are all Nyström Algorithms. In *Proceedings of the 10th International Workshop on Artificial Intelligence and Statistics, Bridgetown, Barbados, 6–8 January 2005*; pp. 261–268.
25. Guttman, L. A general nonmetric technique for finding the smallest coordinate space for a configuration of points. *Psychometrika* **1968**, *33*, 469–506. [CrossRef]
26. Bernataviciene, J.; Dzemyda, G.; Marcinkevicius, V. Diagonal Majorizarion Algorithm: Properties and efficiency. *Inf. Technol. Control* **2007**, *36*, 353–358.
27. Grané, A.; Albarrán, I.; Lumley, R. Visualizing Inequality in Health and Socioeconomic Wellbeing in the EU: Findings from the SHARE Survey. *Int. J. Environ. Res. Public Health* **2020**, *17*, 7747. [CrossRef] [PubMed]
28. Aschenbruck, R.; Szepannek, G. Cluster Validation for Mixed-Type Data. *Achives Data Sci. Ser. A* **2020**. [CrossRef]
29. Foss, A.H.; Markatou, M.; Ray, B. Distance Metrics and Clustering Methods for Mixed-type Data. *Int. Stat. Rev.* **2018**, *81*, 80–109. [CrossRef]
30. Jia, Z.; Song, L. Weighted k-Prototypes Clustering Algorithm Based on the Hybrid Dissimilarity Coefficient. *Math. Probl. Eng.* **2020**, *5143797*. [CrossRef]
31. Paradis, E.; Claude, J.; Strimmer, K. APE: Analyses of phylogenetics and evolution in R language. *Bioinformatics* **2004**, *20*, 289–290. [CrossRef]
32. Dray, S.; Dufour, A.B. The ade4 Package: Implementing the Duality Diagram for Ecologists. *J. Stat. Softw.* **2007**, *22*. [CrossRef]
33. De Leeuw, J.; Mair, P. Multidimensional scaling using majorization: The R package smacof. *J. Stat. Softw.* **2009**, *31*, 1–30. [CrossRef]

34. Oksanen, J.; Blanchet, F.G.; Friendly, M.; Kindt, R.; Legendre, P.; McGlinn, D.; Minchin, P.R.; O'Hara, R.B.; Simpson, G.L.; Solymos, P.; et al. Community Ecology Package, CRAN-Package Vegan. Available online: https://cran.r-project.org; https://github.com/vegandevs/vegan (accessed on 1 March 2020).
35. Roberts, D.W. Ordination and Multivariate Analysis for Ecology. CRAN-Package Labdsv. Available online: http://ecology.msu.montana.edu/labdsv/R (accessed on 1 March 2020).
36. Goslee, S.; Urban, D. Dissimilarity-Based Functions for Ecological Analysis. CRAN-Package Ecodist. Available online: https://CRAN.R-project.org/package=ecodist (accessed on 1 March 2020).
37. Szepannek, G. ClustMixType: User-Friendly Clustering of Mixed-Type Data in R. *R J.* **2018**, *10*, 200–208. [CrossRef]
38. Ney, S. Active Aging Policy in Europe: Between Path Dependency and Path Departure. *Ageing Int.* **2005**, *30*, 325–342. [CrossRef]
39. Avendano, M.; Jürges, H.; MacKenbach, J.P. Educational level and changes in health across Europe: Longitudinal results from SHARE. *J. Eur. Soc. Policy* **2009**, *19*, 301–316. [CrossRef]
40. Bohácek, R.; Crespo, L.; Mira, P.; Pijoan-Mas, J. The Educational Gradient in Life Expectancy in Europe: Preliminary Evidence from SHARE. In *Ageing in Europe—Supporting Policies for an Inclusive Society*; Börsch-Supan, A., Kneip, T., Litwin, H., Myck, M., Weber, G., Eds.; De Gruyter: Berlin, Germany 2015; pp. 321–330. [CrossRef]
41. Sokal, R.R.; Rohlf, F.J. The comparison of dendrograms by objective methods. *Taxon* **1962**, *11*, 33–40. [CrossRef]
42. Grané, A.; Romera, R. On visualizing mixed-type data: A joint metric approach to profile construction and outlier detection. *Sociol. Methods Res.* **2018**, *47*, 207–239. [CrossRef]
43. Grané, A.; Salini, S.; Verdolini, E. Robust multivariate analysis for mixed-type data: Novel algorithm and its practical application in socio-economic research. *Socio-Econ. Plan. Sci.* **2021**, *73*, 100907. [CrossRef]
44. Cuadras, C.M.; Fortiana, J. Visualizing Categorical Data with Related Metric Scaling. In *Visualization of Categorical Data*; Blasius J., Greenacre, M., Eds.; Academic Press: London, UK, 1998; pp. 365–376.
45. Cutler, A.; Breiman, L. Archetypal analysis. *Technometrics* **1994**, *36*, 338–347. [CrossRef]
46. Vinué, G.; Epifanio, I.; Alemany, S. Archetypoids: A new approach to define representative archetypal data. *Comput. Statist. Data Anal.* **2015**, *87*, 102–115. [CrossRef]

Article

Kernel Based Data-Adaptive Support Vector Machines for Multi-Class Classification

Jianli Shao [1,†], Xin Liu [1,2,*,†] and Wenqing He [2,†]

1. School of Statistics and Management, Shanghai University of Finance and Economics, Shanghai 200433, China; jlshao@mail.shufe.edu.cn
2. Department of Statistical and Actuarial Sciences, University of Western Ontario, London, ON N6A 3K7, Canada; whe@stats.uwo.ca
* Correspondence: liu.xin@mail.shufe.edu.cn or xliu246@uwo.ca
† These authors contributed equally to this work.

Abstract: Imbalanced data exist in many classification problems. The classification of imbalanced data has remarkable challenges in machine learning. The support vector machine (SVM) and its variants are popularly used in machine learning among different classifiers thanks to their flexibility and interpretability. However, the performance of SVMs is impacted when the data are imbalanced, which is a typical data structure in the multi-category classification problem. In this paper, we employ the data-adaptive SVM with scaled kernel functions to classify instances for a multi-class population. We propose a multi-class data-dependent kernel function for the SVM by considering class imbalance and the spatial association among instances so that the classification accuracy is enhanced. Simulation studies demonstrate the superb performance of the proposed method, and a real multi-class prostate cancer image dataset is employed as an illustration. Not only does the proposed method outperform the competitor methods in terms of the commonly used accuracy measures such as the F-score and G-means, but also successfully detects more than 60% of instances from the rare class in the real data, while the competitors can only detect less than 20% of the rare class instances. The proposed method will benefit other scientific research fields, such as multiple region boundary detection.

Keywords: classification; data-adaptive kernel functions; image data; multi-category classifier; predictive models; support vector machine

1. Introduction

One of the typical problems in data mining and machine learning is to classify new instances on the basis of observed ones. A common classification problem is separating two classes based on the estimated decision rule trained from the training data, however, multi-class situations have been increasingly seen in various scientific areas, including disease diagnosis in medical research [1], artificial intelligence [2], users' preferences in recommendation systems [3], and risk evaluation in social sciences [4]. Accordingly, techniques are either derived from those binary classifiers or originally proposed specifically for multi-category classification problems. One of the most powerful classifiers is the support vector machine (SVM) [5], which shows its superior performance in many real applications [6] and is known for its excellent performance in both small and big samples, its robustness for outliers, and ease of interpretation.

The most popular framework for dealing with the multi-category classification problems is to decompose it into a series of binary classifications where the regular binary classifiers can be directly applied. Examples of those methods include the well-known **one-versus-one** [7] and **one-versus-all** [5] techniques. In particular, for a k-category classification case under the SVM framework, the least square SVM (LS-SVM) [8] method was extended to the multi-class case [9]. To overcome the drawback of the original LS-SVM that the decision function is constructed from most of the training samples, referred to as the non-sparseness problem, Xia and Li [10] developed a new multi-class LS-SVM algorithm where the solution can be sparse in the weight coefficients of support vectors. Fung and Mangasarian [11] followed the idea of proximal SVM (PSVM) in [12] to extend the PSVM to the multi-class case. For each decomposed sub-classification problem, the solution is similar to its binary case for classifying new samples by allocating them to the closer class of the two parallel planes. This PSVM method turns out to be quite aligned with the one-versus-all method. Zhang et al. [13] extended the PSVM method to include the adaptive kernel function, which magnifies the resolution on each boundary based on weighted factors that can be obtained from a Chi-square distribution. However, its adaptively scaled kernel depends on a squared distance, which may not be reliable [1], and the decay rate for each class is constant. Following the idea by Crammer and Singer [2], He et al. [14] proposed a simplified multi-class support vector machine with a reduced dual optimization. Their method suffers a computation burden. He et al. [14] also presented a simplified multi-class SVM to reduce the size of the resulting dual optimization by introducing a relaxed classification error bound, which speeds up the training process without sacrificing classification accuracy.

However, an imbalance issue usually arises in real applications such as cancer research, especially when dealing with multi-category classification. That is, some minority classes may only contain very few instances in the training sample data when dealing with two categories in nature or using the one-versus-all strategy in multi-class cases. Learning from the imbalanced data turns out to be remarkably challenging in the field of data mining with big data [6]. Many fields have been seeing the importance and need of accurate classifiers for imbalanced data [15], including the detection of rare but serious diseases such as cancers in medical science, fraudulence issues in accounting [16], and risk evaluation in economics [4]. Many commonly used binary classifiers may only show limited predictive power for the minority class when severe imbalances exist [17]. Indeed, this issue corresponds to the unequal distribution of the sample data from different classes, where a majority of instances belong to a specific class while the rest to others. Chawla et al. [18] and Tang et al. [19] have discussed the issue and found that the SVM for multiple classes with imbalanced data can be prone to generating a classifier with a strong estimation bias towards the majority class and will give a rather poor performance. Wang and Shen [20] proposed a method that can avoid the difficulties of the one-versus-all strategy by dealing with multiple classes in a joint manner. Consequently, an accurate classifier is always desired when a specific class is extremely small compared to other classes in the training data, such as the one-versus-all case in dealing with multi-class classification.

To overcome the effect of imbalance on classifications, Liu and He [1] proposed a new method to enhance the performance of the SVM for imbalanced data by adaptively scaling the kernel function obtained from a standard SVM so that the separation between two categories can be effectively enlarged. The method also takes into account the location of the support vectors in the feature space, which makes it more appealing when the responses are from multiple classes. In this paper, we propose a new data-adaptive SVM technique for multi-class problems. A new data-adaptive kernel function is proposed for the multi-class SVM in a way that the decay rate of the scaling magnitude is more robust and can vary along with the density of the samples in the neighborhood. Not only does the method take the imbalance of data from a multi-class response into consideration, but it involves spatial association of local data instances as well. By using this adaptive kernel function, the constructed classifier shows excellent predictive power, especially for imbalanced data, with a competitive cost of time consumption. Numerical investigations demonstrate the superior performance of the proposed method, and a real image dataset is employed as an illustration.

The remainder of the paper is organized as follows. Section 2 introduces the proposed methodology for multi-category classification with class imbalance taken into account. Numerical investigation is presented in Section 3 to demonstrate the superb prediction accuracy of the proposed method compared with its competitors. Concluding remarks and discussion are described in the final section.

2. Methodology

2.1. SVM Framework and Notation

As a general method for classification proposed by Vapnik and Vapnik [5], the support vector machine essentially uses a kernel function that maps the original input data space into a high-dimensional feature space so that the instances from two classes are as far as possible, preferably separable with a linear boundary in the feature space.

To start with, we consider a binary case. Given a sample $\{x_i, y_i\}$ for $i = 1, \ldots, n$, where x_i is a vector of predictors in the input space $I = R^p$ and y_i represents the class index, which takes a value from $\{+1, -1\}$, a nonlinear support vector machine maps the input data $x = \{x_1, \ldots, x_n\}$ into a high-dimensional feature space, $F = R^l$, using a nonlinear mapping function $s : R^p \to R^l$, and finds a linear boundary in the feature space F by maximizing the smallest distance of instances to this boundary. Mathematically, the idea is equivalent to solve

$$\min_{w,b} \quad \frac{1}{2} w^T w + C \sum_{i=1}^{n} \xi_{i,t} \tag{1}$$

$$\text{subject to} \quad y_i(w^T s(x_i) + b) \geq 1 - \xi_{i,t},$$
$$\xi_{i,t} \geq 0, \quad \text{for } i = 1, \ldots, n,$$

where C is the so-called soft margin parameter that determines the trade-off between the optimal combinatorial choice of the margin and the classification error, and $\xi = (\xi_1, \ldots, \xi_n)^T$ is a non-negative slack variable vector that controls misclassification. The dual procedure of (1) is to solve

$$\text{Max}_{\alpha} \quad \sum_{i=1}^{n} \alpha_i - \frac{1}{2} \sum_{i=1}^{n} \sum_{j=1}^{n} \alpha_i \alpha_j y_i y_j K(x_i, x_j), \tag{2}$$

$$\text{subject to} \quad \sum_{i=1}^{n} \alpha_i y_i = 0, \ 0 \leq \alpha_i \leq C, \ i = 1, 2, \ldots, n, \tag{3}$$

where α_i's are the dual variables and the scalar function $K(\cdot, \cdot)$ is called a kernel function defined as $K(x_i, x_j) = <s(x_i), s(x_j)>$ with $<\cdot, \cdot>$ being the inner product operator. Denote SV the index set of the support vectors $\{j \mid \alpha_j > 0 \text{ for } j = 1, 2, \ldots, n\}$. With all the observations $x_i, i \in SV$, the kernel form of the SVM boundary can be written as

$$\sum_{i \in SV} \alpha_i y_i K(x_i, x) + b = 0. \tag{4}$$

Consequently, the label of an instance x is assigned by $sign(D(x))$, with

$$D(x) = \sum_{i \in SV} \hat{\alpha}_i y_i K(x_i, x) + \hat{b}. \tag{5}$$

where $\hat{a}$ represents the predicted value of a. Theoretically, the bias term b_j is proved identical for all instances in the SV [21]. Practically, the biased term $\hat{b}$ is determined as the average of all the estimated $\hat{b}_j$'s at all the support vectors, where $\hat{b}_j$ is obtained by using the j-th support vector x_j

$$\hat{b}_j = y_j - \sum_{i \in SV} \hat{\alpha}_i y_i K(x_i, x_j).$$

A k-category classification problem with the class label y_i taking value from $\{1, \ldots, k\}$ can be generally decomposed into a sequence of binary classification problems using the one-versus-all strategy. Specifically, the m-th binary classification, $m = 1, \ldots, k$, is set up for a training sample $\{x_i, y_i^{(m)}\}$, where $y_i^{(m)} = I(y_i = m) - I(y_i \neq m)$ and $I(\cdot)$ is the indicator function. Hence, by applying the SVM procedure for binary classification, k classifiers can be constructed with k kernels $K_1, \ldots, K_k$, and the m-th kernel form of the SVM boundary between the m-th class and the remaining $(k-1)$ classes can be written as

$$D_m(\mathbf{x}) = \sum_{i \in SV_m} \alpha_i^{(m)} y_i^{(m)} K_m(\mathbf{x}_i, \mathbf{x}) + b^{(m)} \tag{6}$$

With the estimated decision functions from all m-th binary classifications, the final class label of an instance can be assigned using a majority voting procedure.

Quite a few typical kernels are available for the SVM procedure. One is the radial kernel $K(\mathbf{x}, \mathbf{x}') = f(-\|\mathbf{x} - \mathbf{x}'\|^2/2)$, such as the Gaussian Radial Basis Function kernel,

$$K(\mathbf{x}, \mathbf{x}') = \exp(-\|\mathbf{x} - \mathbf{x}'\|^2/2\sigma^2).$$

Another type of kernel takes a form of the inner product $K(\mathbf{x}, \mathbf{x}') = f(<\mathbf{x}, \mathbf{x}'>)$, such as a polynomial kernel with degree d,

$$K(\mathbf{x}, \mathbf{x}') = (1 + <\mathbf{x}, \mathbf{x}'>)^d.$$

2.2. Conformal Transformation and Adaptive Kernel Machine

From the geometrical point of view, when the feature space F is the Euclidean space, the Riemannian metric is induced in the input space I. Take a two-dimensional case, for instance, a small change $\mathbf{d}(\mathbf{x})$ in the input space will be mapped as $\mathbf{ds}(\mathbf{x})$ in the feature space

$$\mathbf{ds}(\mathbf{x}) = \nabla \mathbf{s} \cdot \mathbf{dx}, \tag{7}$$

where

$$\nabla \mathbf{s} = \left(\frac{\partial \mathbf{s}(\mathbf{x})}{\partial \mathbf{x}} \right) = \begin{pmatrix} \frac{\partial s_1(\mathbf{x})}{\partial x_1} & \cdots & \frac{\partial s_1(\mathbf{x})}{\partial x_p} \\ \vdots & \vdots & \vdots \\ \frac{\partial s_l(\mathbf{x})}{\partial x_1} & \cdots & \frac{\partial s_l(\mathbf{x})}{\partial x_p} \end{pmatrix}. \tag{8}$$

Thus, the squared length of $\mathbf{ds}(\mathbf{x})$ can be written in the quadratic form as

$$\|\mathbf{ds}(\mathbf{x})\|^2 = (\mathbf{ds}(\mathbf{x}))^T \mathbf{ds}(\mathbf{x}) = \sum_{ij} s_{ij}(\mathbf{x}) dx_i \, dx_j, \tag{9}$$

where

$$s_{ij}(\mathbf{x}) = (\nabla \mathbf{s})^T \cdot (\nabla \mathbf{s}). \tag{10}$$

Lemma 1 ([1]). *Suppose $K(p, q)$ is a kernel function, and $s(\cdot)$ is the corresponding mapping in the support vector machine. Then*

$$s_{ij}(p) = \frac{\partial}{\partial p_i} \frac{\partial}{\partial q_j} K(p, q)|_{q=p}. \tag{11}$$

Detailed proof is given in Appendix A.

Though the parameters of kernel functions are able to manipulate the geometric characteristics of the feature space F to some degree, conformal transformation on the original kernel function can further contribute to great adaptability. Conformal transformation is a function mapping that projects the original input space to a new feature space with the angles between vectors being preserved in a local area [1]. Define

$$\tilde{\mathbf{s}}(\mathbf{x}) = c(\mathbf{x})\mathbf{s}(\mathbf{x}) \tag{12}$$

and
$$\tilde{K}(\mathbf{x},\mathbf{x}') =<\tilde{\mathbf{s}}(\mathbf{x}),\tilde{\mathbf{s}}(\mathbf{x})'>= c(\mathbf{x})c(\mathbf{x}')<\mathbf{s}(\mathbf{x}),\mathbf{s}(\mathbf{x}')>= c(\mathbf{x})c(\mathbf{x}')K(\mathbf{x},\mathbf{x}'), \quad (13)$$

then $\tilde{K}(\mathbf{x},\mathbf{x}')$ corresponds to the mapping $\tilde{\mathbf{s}}$ that may increase the separation for a properly chosen positive scalar function $c(\mathbf{x})$ which has larger values at the support vectors identified using the kernel $K(\mathbf{x},\mathbf{x}')$. Furthermore, $\tilde{K}$ can be easily shown to satisfy the Mercer positivity condition, the sufficient condition for being a kernel function. Specifically, we employ the L_1-norm adaptive radial basis function (RBF) kernel proposed in [1]:

$$c(\mathbf{x}) = e^{-|D(\mathbf{x})|d_M(\mathbf{x})} \quad (14)$$

where
$$d_M(\mathbf{x}) = AVG_{i \in \{\|\mathbf{s}(\mathbf{x}_i) - \mathbf{s}(\mathbf{x})\|^2 < M, \, y_i \neq y\}}(\|\mathbf{s}(\mathbf{x}_i) - \mathbf{s}(\mathbf{x})\|^2), \quad (15)$$

and M can be regarded as the distance between the nearest and the farthest support vectors under the original mapping $\mathbf{s}(\mathbf{x})$. In this way, the average on the right-hand side can comprise all the support vectors different from the currently considered instance in the neighborhood of $\mathbf{s}(\mathbf{x})$ within the radius of M. This takes into account the spatial distribution of the support vectors in the feature space F, and hence partially reflects the spatial association of the instances in the training set. This method turns out to be robust and efficient [1].

2.3. Adaptive Kernel Machine for Multi-Class Cases

To apply the adaptive kernel machine to a multi-class classification problem, we first apply the basic SVM to all k classes of a training sample by employing the one-versus-all strategy, and obtain k initial decision boundaries as well as the predicted labels of all instances. We then split the training sample in k datasets using the label of class $\hat{y}_i$ from the initial round SVM, represented by $S_1, S_2, \ldots, S_k$, respectively. This step is essential in the sense of finding the approximated locations of support vectors and the initial boundaries. Similar to the idea of conformal transformation in the binary case, the adaptive data-dependent kernel transformation function is defined as

$$c(\mathbf{x}) = \begin{cases} \exp(-p_1(\mathbf{x})|D_1(\mathbf{x})|), & \text{if } \mathbf{x} \in S_1 \\ \exp(-p_2(\mathbf{x})|D_2(\mathbf{x})|), & \text{if } \mathbf{x} \in S_2 \\ \ldots \\ \exp(-p_k(\mathbf{x})|D_k(\mathbf{x})|), & \text{if } \mathbf{x} \in S_k \end{cases} \quad (16)$$

where $p_m(\mathbf{x})$, $m = 1, \ldots, k$, are functions of data that will be determined to control the decay rates and hence further affect the performance of the classifier.

2.4. Specification of Functions $p_m(\mathbf{x})$

In an imbalanced data classification, determination of appropriate weights for each category is important so that the problem can be transferred back to the approximately balanced case. Generally, there are two requirements for the choice of weights. One is that the data in the majority class should be allocated with a smaller weight than those in the minority class so that the data are somewhat balanced in the contribution to the decision function. The other is the natural restriction that the sum of the weights should be 1. Essentially, for imbalanced data, the weights can be set as the reciprocal of the sizes of the classes in the training sample. Let n_m denote the training sample size for the m-th class, $m = 1, \ldots, k$. Then the weightings are defined as

$$w_m = \frac{1/n_m^2}{\sum_{i=1}^{k} 1/n_m^2}. \quad (17)$$

In this way, w_ms show the sparse distribution nature of each category. Note that a L_2-norm is adopted when building w_m in (17). Although L_p-norm ($p > 0$) can be applied in general, such as the L_1-norm, in real applications, we found the L_2-norm would show the best empirical performance.

As w_ms do not involve the information of $\mathbf{x}$, we further introduce the idea of constructing $c(\mathbf{x})$ in the binary case to include information from $\mathbf{x}$. Define

$$d_m(\mathbf{x}) = AVG_{j \in SV_m}(\|\mathbf{s}_m(\mathbf{x}_j) - \mathbf{s}_m(\mathbf{x})\|^2) = AVG_{j \in SV_m}(K_m(\mathbf{x}_j, \mathbf{x}_j) + K_m(\mathbf{x}, \mathbf{x}) - 2K_m(\mathbf{x}_j, \mathbf{x})) \quad (18)$$

where SV_m is the support vector set from the initial SVM with the binary SVM procedure in the m-th class, $K_m(\cdot, \cdot)$ is the kernel function adopted in the m-th binary SVM and $\mathbf{s}_m(\cdot)$ is its corresponding mapping function. In practice, we adopt a common kernel function $K_m(\cdot, \cdot), m = 1, \ldots, k$, such as the popular Gaussian kernel function, to simplify the calculation. Consequently, we define

$$p_m(\mathbf{x}) = w_m \cdot d_m(\mathbf{x}) \quad (19)$$

so that the influence from the size of the class is taken into account.

Another potential choice of $p_m(\mathbf{x})$ could be

$$p_m(\mathbf{x}) = AVG_{i \in \{\|\mathbf{s}_m(\mathbf{x}_i) - \mathbf{s}_m(\mathbf{x})\|^2 < Q_m,\ y_i \neq y\}}(\|\mathbf{s}_m(\mathbf{x}_i) - \mathbf{s}_m(\mathbf{x})\|^2), \quad (20)$$

where the tuning parameter Q_m can also be regarded as the distance between the nearest and the farthest support vector in SV_m from $\mathbf{s}_m(\mathbf{x})$ within the same class. When k is small or moderate, this setting can be meaningful. However, when k is large, the computational cost may arise since more tuning parameters need to be determined. To avoid the problem, we propose to use a universal control Q while taking the weights w_m into account. The final version of $p_m(\mathbf{x})$ is constructed as

$$p_m(\mathbf{x}) = AVG_{i \in \{\|\mathbf{s}_m(\mathbf{x}_i) - \mathbf{s}_m(\mathbf{x})\|^2 < Q \cdot w_m,\ y_i \neq y\}}(\|\mathbf{s}_m(\mathbf{x}_i) - \mathbf{s}_m(\mathbf{x})\|^2) \quad (21)$$

In this way, the classification can be more robust to extreme cases in spatial distribution, which may push the classification boundaries towards the majority classes, while the weights are considered to balance the training set so that the performance of the classification is enhanced.

Some other techniques are seen in the literature, though they may show some drawbacks in different situations for imbalanced data. For example, Wu and Amari [22] made some improvements by introducing different tuning parameters for different classes so that the local density of support vectors can be accommodated. With the heavy computational cost it brings, the performance in high-dimension cases turns out uncertain. Williams et al. [23] also extended their binary scaling SVM technique to the multi-class case; however, its distance tuning parameter, corresponding to the value of $Q \cdot w_m$ in our case, is fixed throughout the whole region. This inflexible setting cannot reflect the local information, especially when the density of support vectors is quite high. Also, using L_2-norm of $D(\mathbf{x})$ may lead to unstable classification performance in high dimensional cases due to a faster decay rate to a constant e^{-k} compared with our proposed method.

2.5. Data-Adaptive SVM Algorithm for Multi-Class Case

With $c(\mathbf{x})$ constructed in (16), we conformally transfer the k kernels trained from the initial round of multi-class SVM, $K_1, \ldots, K_k$ into

$$\tilde{K}_m(\mathbf{x}_i, \mathbf{x}_j) = c(\mathbf{x}_i)c(\mathbf{x}_j)K(\mathbf{x}_i, \mathbf{x}_j), \quad (22)$$

where $m = 1, \ldots, k$, $c(\cdot)$ is defined in (16) with $p_m(\mathbf{x})$ as (21). $K_m(\cdot, \cdot)$ is usually set as the Gaussian kernel function during the first round of SVM. The performance of using the form in (19) is similar empirically. Based on the updated kernels, the second round SVM is then conducted and predictions of labels for all instances are obtained. It is seen that

1. The magnification will be almost constant along the separating surface $D(\mathbf{x}) = 0$ for each boundary;
2. The magnification will be largest where the contours are closest locally. (See more details in the Appendices.)

Thus, as long as the parameters C and σ in the kernel machine (and the controlling parameter Q if the form of $p_m(\mathbf{x})$ in (21) is adopted) are tuning adaptively with data, the classifiers can be trained, and hence the subjects' labels can be predicted.

To conclude the section, the algorithm of the whole procedure of the multi-label classification problem is described as follows. A regular SVM classifier is trained with an ordinary Gaussian radial basis kernel function, and the support vectors are found so that the separating boundaries can be approximately determined using the one-versus-all technique in the first stage. Based on the spatial information of the support vectors, the conformal transformations will be constructed, and the original kernel functions are updated. Then a new round of SVM optimization problems is conducted with the updated kernel function so that the boundary in each one-versus-all strategy can be found. Consequently, the predicted labels for subjects can be estimated. The whole procedure is summarized in Algorithm 1.

Algorithm 1. Multi-class data adaptive kernel scaling support vector machine (SVM).

Input: $y_i, \mathbf{x}_i, i = 1, \ldots, n$; a Gaussian kernel function $K(\cdot, \cdot)$
1: A regular SVM classifier is trained with an ordinary Gaussian radial basis kernel function;
2: Based on the spatial information of these support vectors, the conformal transformation is constructed, and the original kernel function is updated;
3: A new round of SVM optimization problems is conducted with the updated kernel function, and the boundaries for different classes are found;
4: The predicted class labels for instances are determined by majority voting.

---

3. Numerical Investigation

In this section, we conduct intensive numerical experiments to evaluate the performance of the proposed classification procedure and compare them with the existing competitors. The whole study will be divided into two parts, one for simulated data and the other for a real image dataset. We will compare the proposed method with four existing methods, including the traditional SVM and methods from Wu [22], William [23] and Maratea [24].

We assess the performance of the classifiers using various quantitative measures. One of them is the overall accuracy, defined by

$$P_{overall} = \frac{TP+TN}{TP+FP+TN+FN} = \frac{TP+TN}{n},$$

where TP, FN, FP and TN represent the number of instances of true positive, false negative, false positive and true negative in the test sample, respectively. However, for imbalanced data, the overall accuracy rate may not be sufficient [24]. We further adopt two other measurements on classifiers' performance for imbalanced data, namely the F-score and the G-mean, respectively [25]. Specifically, the F-score is defined as

$$F_{score} = \frac{2 \times P_{pre} \times P_{spe}}{P_{pre} + P_{sen}},$$

and G-mean as

$$G_{mean} = \sqrt{P_{sen} \times P_{spe}},$$

where P_{pre}, P_{sen} and P_{spe} are the precision, the sensitivity and the specificity, respectively. They are obtained by

$$P_{pre} = \frac{TP}{TP+FP},$$

$$P_{sen} = \frac{TP}{TP+FN},$$

and

$$P_{spe} = \frac{TN}{TN+FP}.$$

Note that F-score measures the harmonic mean of the precision and sensitivity, while G-mean is constructed as the geometric mean of the sensitivity and the specificity, giving a more fair comparison between the positive and negative classes, regardless of its size. To further evaluate the numerical performance of the multi-category classification, we employ the multi-class ROC and the AUC measures [25].

3.1. Simulation Study

First, we conduct simulation studies to evaluate the performance of the proposed method and compare it with the competitors in the literature. Three scenarios are considered. Each of them includes the balanced, moderately imbalanced, and extremely imbalanced cases, respectively. The Gaussian RBF kernel is employed during the first round of classification, if not mentioned elsewhere.

For convenience, the input space is 2-dimensional, and all training data are generated using three classes of bivariate Gaussian distributions with means vectors $(2,2)$, $(4,3)$, $(3,2)$, and identical covariance matrix $\gamma \cdot \Sigma$, where γ is a nuisance parameter that controls the overlapping proportion of the classes. Moderate covariance is incorporated for all pairs with a correlation coefficient $\rho = 0.3$, and the variance of all variables is 1.

The overall sample size for the training data is set as 600 and is separated into three classes by different weights in three different scenarios. The class size is $(200, 200, 200)$ in Scenario 1, $(100, 200, 300)$ in Scenario 2, and $(20, 100, 480)$ in Scenario 3. In each scenario, different combinations of parameters that need to be tuned will be considered. The cost parameter C is chosen from the set $\{0.1, 0.2, 0.5, 1, 5, 8, 40, 100, 500\}$ and σ takes value from the set $\{0.01, 0.05, 0.1, 0.5, 1, 5, 10, 100\}$. As Q is the threshold controlling the size of the local neighborhood, it is chosen by a grid search from the set $\{0.1, 0.2, \ldots, 1\}$ times the maximal Euclidean distance between all pairs of data points in the sample. All classifiers are tuned properly with respect to the corresponding measures.

The classification procedure is as follows. First, we train the classifiers with the traditional SVM using the one-versus-all strategy, and the support vectors are identified approximately. The kernel functions for all the methods are then updated adaptively by conformal transformation with different scalar function $c(\mathbf{x})$, using p_m defined in (21). A second round of SVM is then conducted, and the estimated class labels for observations in the test sample will be given and consequently compared with the true labels. Five-fold cross validation is employed to obtain the misclassification rate for each simulated dataset, and the whole process is repeated 1000 times. With the accuracy measures defined above, the performance of all the classifiers is shown in Tables 1–3, and Figures 1–3. Similar results are seen in the proposed method with the other way of defining p_m.

Table 1. F-score (F), G-mean (G) and the AUC (A) measures for all five classification methods in Scenario 1 for $n_1 = 200$, $n_3 = 200$ and $n_3 = 200$, respectively. Max margin is 0.02.

		SVM			Wu [22]			William [23]			Maratea [24]			Our Method		
C	σ	F	G	A	F	G	A	F	G	A	F	G	A	F	G	A
8	0.1	0.39	0.38	0.52	0.43	0.43	0.54	0.59	0.59	0.61	0.71	0.70	0.75	0.78	0.79	0.80
8	0.5	0.43	0.42	0.55	0.47	0.46	0.56	0.66	0.66	0.69	0.75	0.75	0.78	0.81	0.81	0.83
8	5.0	0.47	0.46	0.56	0.53	0.52	0.59	0.68	0.68	0.72	0.78	0.77	0.81	0.84	0.83	0.85
40	0.1	0.45	0.45	0.55	0.44	0.43	0.54	0.61	0.61	0.66	0.73	0.72	0.75	0.81	0.81	0.85
40	0.5	0.53	0.52	0.57	0.51	0.50	0.55	0.67	0.67	0.71	0.75	0.75	0.78	0.84	0.83	0.88
40	5.0	0.56	0.55	0.59	0.62	0.62	0.67	0.71	0.72	0.78	0.78	0.78	0.81	0.86	0.86	0.88
100	0.1	0.52	0.51	0.57	0.61	0.59	0.62	0.64	0.63	0.68	0.78	0.79	0.81	0.84	0.85	0.88
100	0.5	0.60	0.58	0.62	0.67	0.65	0.69	0.77	0.67	0.76	0.79	0.80	0.83	0.86	0.86	0.90
100	5.0	0.69	0.66	0.71	0.71	0.70	0.73	0.79	0.72	0.80	0.81	0.82	0.84	0.88	0.88	0.92

Table 2. F-score (F), G-mean (G) and the AUC (A) measures for all five classification methods in Scenario 2 for $n_1 = 100$, $n_3 = 200$ and $n_3 = 300$, respectively. Max margin is 0.04.

		SVM			Wu [22]			William [23]			Maratea [24]			Our Method		
C	σ	F	G	A	F	G	A	F	G	A	F	G	A	F	G	A
8	0.1	0.29	0.30	0.51	0.40	0.40	0.52	0.49	0.49	0.55	0.62	0.60	0.63	0.77	0.76	0.80
8	0.5	0.32	0.32	0.51	0.42	0.42	0.55	0.58	0.57	0.63	0.66	0.65	0.71	0.78	0.78	0.82
8	5.0	0.36	0.36	0.52	0.48	0.49	0.55	0.61	0.60	0.67	0.71	0.72	0.77	0.80	0.80	0.84
40	0.1	0.40	0.40	0.54	0.54	0.53	0.57	0.57	0.58	0.62	0.65	0.66	0.71	0.80	0.80	0.83
40	0.5	0.50	0.42	0.52	0.59	0.58	0.64	0.65	0.64	0.69	0.68	0.67	0.73	0.81	0.80	0.85
40	5.0	0.56	0.56	0.61	0.62	0.60	0.64	0.68	0.66	0.72	0.72	0.73	0.77	0.82	0.82	0.88
100	0.1	0.42	0.39	0.54	0.57	0.55	0.61	0.65	0.64	0.69	0.66	0.68	0.75	0.84	0.83	0.88
100	0.5	0.52	0.50	0.59	0.63	0.65	0.71	0.70	0.69	0.75	0.71	0.72	0.77	0.85	0.84	0.89
100	5.0	0.59	0.61	0.66	0.68	0.70	0.74	0.75	0.76	0.79	0.75	0.74	0.81	0.86	0.87	0.91

Table 3. F-score (F), G-mean (G) and the AUC (A) measures for all five classification methods in Scenario 3 for $n_1 = 20$, $n_3 = 100$ and $n_3 = 480$, respectively. Max margin is 0.05.

		SVM			Wu [22]			William [23]			Maratea [24]			Our Method		
C	σ	F	G	A	F	G	A	F	G	A	F	G	A	F	G	A
8	0.1	0.25	0.23	0.50	0.34	0.32	0.51	0.45	0.46	0.53	0.58	0.57	0.61	0.75	0.74	0.79
8	0.5	0.28	0.27	0.51	0.37	0.38	0.53	0.54	0.52	0.60	0.62	0.64	0.69	0.77	0.77	0.82
8	5.0	0.32	0.29	0.51	0.46	0.44	0.53	0.57	0.58	0.62	0.68	0.66	0.72	0.79	0.79	0.84
40	0.1	0.35	0.34	0.51	0.51	0.49	0.54	0.53	0.54	0.59	0.60	0.59	0.64	0.79	0.78	0.84
40	0.5	0.47	0.45	0.54	0.55	0.54	0.60	0.59	0.58	0.63	0.64	0.64	0.71	0.80	0.80	0.85
40	5.0	0.51	0.52	0.58	0.57	0.55	0.62	0.63	0.62	0.68	0.68	0.69	0.75	0.81	0.81	0.84
100	0.1	0.38	0.37	0.51	0.54	0.52	0.58	0.61	0.60	0.64	0.62	0.63	0.70	0.82	0.82	0.85
100	0.5	0.45	0.45	0.55	0.59	0.57	0.64	0.65	0.64	0.72	0.67	0.67	0.74	0.84	0.84	0.87
100	5.0	0.55	0.55	0.60	0.62	0.63	0.68	0.71	0.71	0.76	0.71	0.70	0.75	0.85	0.86	0.91

It is seen that all methods considered here have improved performances comparing to the ordinary SVM in almost all scenarios with different combinations of the parameters C and σ. In general, the proposed method outperforms all the other classifiers considered, especially in the imbalanced data. When σ gets larger with fixed C, the misclassification rate tends to decrease in all the methods compared. When σ is relatively small, the proposed method performs better than those of Wu and Williams' methods, while if σ is relatively large, all the methods are nearly the same. This is because when σ is large, the feasible solution set gets large, and all of the methods tend to find the optimal solution. Correspondingly, when C increases, the budget for misclassification gets bigger, which means more tolerance is permitted so that the two classes can be separated. In this scenario, we found that p_m is roughly the same, approximately the reciprocal of $|D|_{max}$. This makes sense because in the balanced-data case, the density of the distributed SVs is roughly uniform, and hence the averages of the distance in the feature space for each data point are roughly the same.

For imbalanced scenarios the performances of all methods turn out to be a bit worse than the balanced case with no surprise due to the non-uniformly distributed support vectors. The change of the misclassification rate with C and σ is similar to that in the balanced data case. The proposed method performs the best among all the methods.

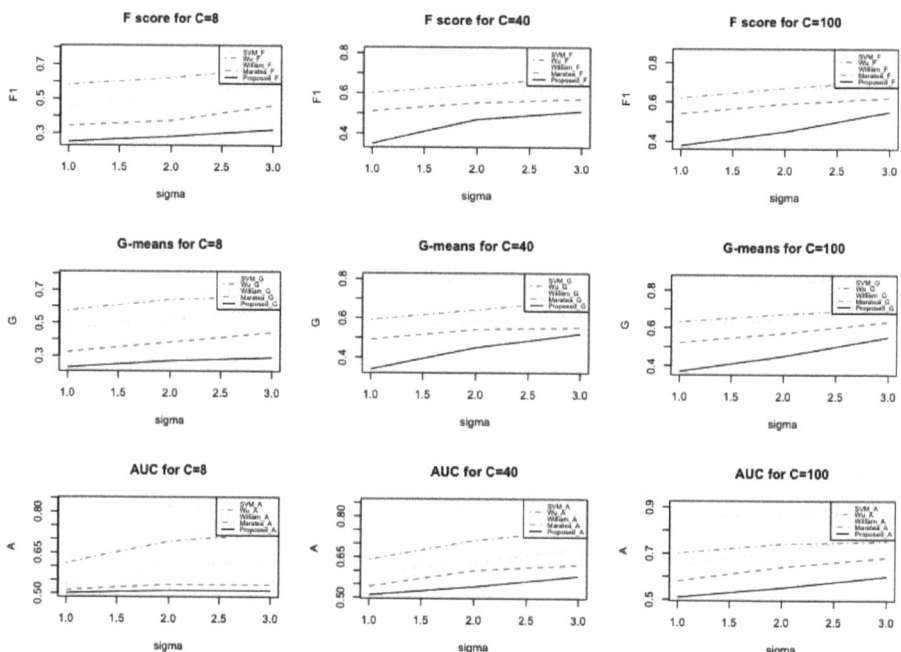

Figure 1. Performance of the competitor methods for Scenario 1.

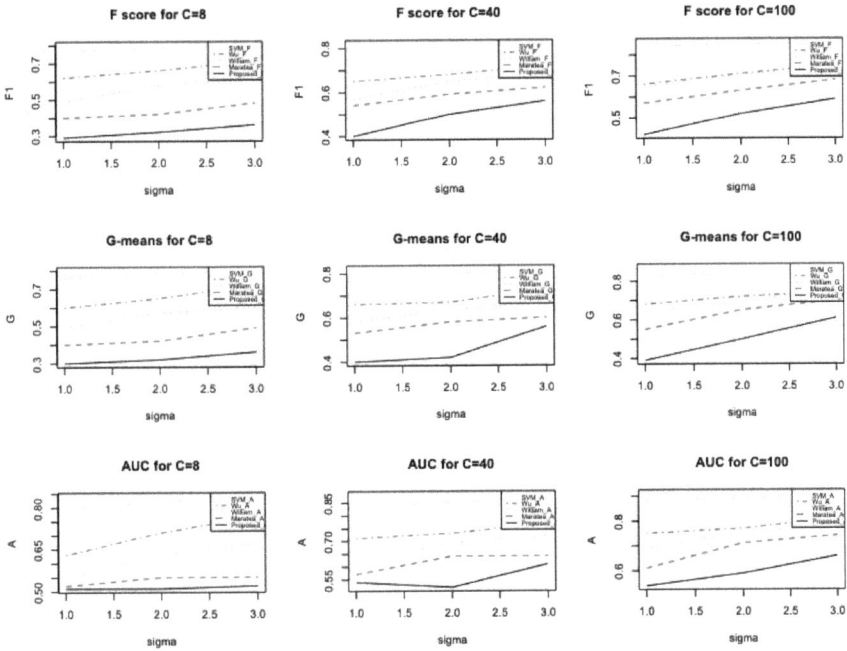

Figure 2. Performance of the competitor methods for Scenario 2.

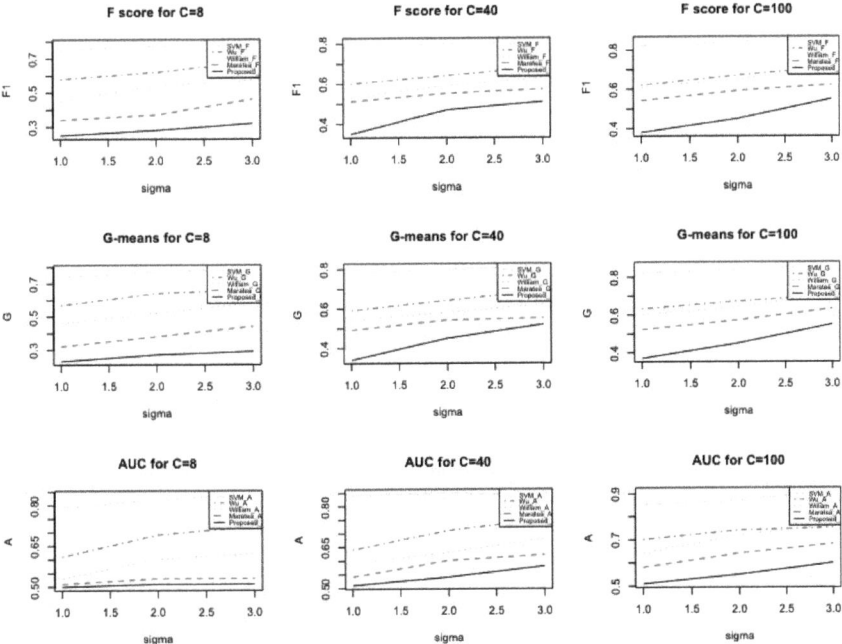

Figure 3. Performance of the competitor methods for Scenario 3.

3.2. A Real Prostate Caner MRI Dataset

In this section, we apply the proposed method to a prostate cancer MR image dataset. The study aims to find statistical methods to classify cancer and non-cancer areas or grades of cancer by the imaging data obtained by imaging collection equipments. In this case, nine common classes are labeled and listed as follows, indicating different levels of severity of cancer.

- Atrophy: As means literally (non-cancer);
- EPE: Prostatic intraepithelial neoplasia (non-cancer);
- PIN: Prostatic intraepithelial neoplasia (non-cancer);
- G3: Tumour focus that is all Gleason 3 (cancer);
- G4: Tumour focus that is all Gleason 4 (cancer);
- G3+4: Tumour focus all predominately G3 with intermingled G4 (cancer);
- G4+3: Tumour focus all predominately G4 with intermingled G3 (cancer);
- G4+5: Tumour focus all predominately G4 with intermingled G5 (cancer);
- OtherProstate: Prostate tissue that does not fall into the other categories (non-cancer).

Note that the labels are given at the voxel level. That is, for a specific patient, it is very likely to have different voxels (indicating different positions of the prostate tissue) with different classes. A patient that has $G3+4$-type cancer in some areas is likely to have $G3$-type cancer as well as *OtherProstate*-type of voxels in other areas. Our objective is to predict class labels at the voxel level. There are several labels associated with $G5$, however, the whole dataset contains only one patient with a very tiny area of $G5$ and the associated type of cancer. Therefore, $G5$ is extremely imbalanced.

In the first phase of the study, 21 patients are involved and more than 400 images are collected. Predictors on each voxel are the three-dimensional intensity measures from MRIs, denoted as T2W intensity, ADC intensity and C-Grade intensity. Other measures such as DCE and DWI are only available to part of the patients and hence are not included in the training process.

To adopt the proposed data-adaptive scaling in this multi-class case, two-stage SVMs are required. During the first stage, a standard SVM with the selected kernel is conducted so that the support vectors from the original dataset can be found. Based on the identified support vectors, the kernel functions are updated. Then, a second-stage SVM is conducted with the updated kernel, and the resulting estimated boundary will be used as the rule for classification. In terms of choosing appropriate tuning parameters for each method, 7-fold cross validation is conducted for 500 times at the patient level.

To assess the performance, we compare our proposed methods with both traditional and data-adaptive multi-category classification methods. In terms of the traditional methods, one-versus-one (1vs1) and one-versus-all (1vsA) from indirect methods, and the Crammer and Singer's (CS) direct methods and He's Simplified SVM (simSVP) will be included, while for the data-adaptive methods, Amari's and William's adaptively scaling will be included. In terms of the criterion of the classification performance, misclassification rate, percentage of support vectors in the whole dataset, F-score and G-means along with their margins are reported.

Table 4 presents the assessment measures for all the methods considered. Obviously, the proposed method performs almost the best among all the compared methods. A highlight point is that the proposed method has the smallest margins in all performance measures, resulting from the property of the robust decay of the magnification effect from the proposed data-adaptive kernel. In terms of the accuracy, the proposed method has a similar misclassification rate to the indirect methods, which is significantly smaller than the rest of the methods. F-score and G-means are the largest for the proposed method, much larger than other data-adaptive kernel methods. The percentage of support vectors that are used for constructing the classifiers is the smallest for the proposed method.

Table 4. Outcomes of multi-class prediction on the Prostate Cancer Program.

Methods	Error(%)	SV(%)	F-Score	G-Means
Proposed	8.06 ± 0.58	17.46 ± 0.57	0.84 ± 0.05	0.81 ± 0.04
Amari	11.88 ± 1.12	21.33 ± 1.27	0.70 ± 0.09	0.66 ± 0.10
William	10.21 ± 0.97	18.93 ± 1.65	0.74 ± 0.12	0.71 ± 0.08
CS	9.20 ± 1.22	17.57 ± 1.12	0.77 ± 0.06	0.73 ± 0.06
simSVM	9.33 ± 1.20	18.29 ± 1.07	0.78 ± 0.10	0.74 ± 0.09
1vs1	8.20 ± 1.26	25.41 ± 2.87	0.81 ± 0.06	0.77 ± 0.07
1vsA	8.25 ± 1.57	24.16 ± 2.62	0.82 ± 0.06	0.76 ± 0.06

It is worth pointing out that among those wrongly predicted labels, $G4 + 3$ is the dominant class. In other words, the misclassification always happens in $G4 + 3$ type cancer. This is because this type of cancer is really rare in the training sample, taking only 1–2% among all the labels. These extremely imbalanced data have made it very difficult to be detected with a high accuracy. The proposed method can detect around 60% among this type, while other data adaptive (Amari's and William's) methods can only find less than 20%. All other methods cannot detect this class. Also, only our method detects the $G5$ class from the only one patient, while all competitor methods fail.

4. Concluding Remarks

In this paper, we developed a new data-dependent SVM construction technique for the multi-category classification problem. Based on the data-adaptive kernel SVM for the binary case, we proposed a new method to construct the data-dependent kernel for the multi-class setting, especially when the data are imbalanced. The data-dependent kernel functions have a more robust decay rate and can vary along with the density of the size of neighbors. Thus, the kernel can be adapted optimally for a specific dataset. Numerical results from both synthetic and real datasets have shown the excellent performance of the proposed method. Not only does the proposed method outperform in terms of the commonly used accuracy measures such as the F-score and G-means, compared with the competitors, but also successfully detects more than 60% of instances from the rare class in the real data, while the competitors can only detect less than 20%. A possible future work is to select relevant predictors for the multi-class kernel functions and consider the spatial association between different images. It is worth noting that the misclassification rate will be affected by the distance of the mean vectors. For instance, the misclassification will not occur if the centers of the three Gaussian distributions are sufficiently far from each other when the covariance matrix is set as unity. The proposed method may be useful in other scientific research fields, such as detecting the boundaries of multiple regions of interest.

Author Contributions: Conceptualization, J.S.; Formal analysis, X.L.; Funding acquisition, W.H.; Investigation, W.H.; Methodology, X.L.; Supervision, W.H.; Validation, X.L.; Writing—original draft, J.S. and X.L.; Writing—review and editing, J.S., X.L. and W.H. All authors have read and agreed to the published version of the manuscript.

Funding: This research received no external funding.

Institutional Review Board Statement: Not applicable.

Informed Consent Statement: Not applicable.

Data Availability Statement: Not applicable.

Acknowledgments: Liu's research was partially supported by the Fundamental Research Funds for the Central Universities. He's research was partially supported by the Natural Science and Engineering Research Council of Canada (NSERC). The authors thank the CIHR Team at Image-Guided Prostate Cancer Management at the University of Western Ontario. The authors also thank the reviewer team for their constructive comments.

Conflicts of Interest: The authors declare no conflict of interest.

Appendix A. Proof of Lemmas and Theorems

Appendix A.1. Proof of Lemma 1

Proof. By the definition of a reproducing kernel function $K(\mathbf{x}, \mathbf{z})$ with its values λ_k and the corresponding scalar eigenfunctions $g_k(\mathbf{x})$, we have

$$\int K(\mathbf{x}, \mathbf{z}) \cdot g_k(\mathbf{z}) \, d\mathbf{z} = \lambda_k \cdot g_k(\mathbf{x})$$

where $k = 1, 2, \ldots, l$. Then the kernel is represented as

$$K(\mathbf{x}, \mathbf{z}) = \sum_k \lambda_k \cdot g_k(\mathbf{x}) \cdot g_k(\mathbf{z}).$$

By rescaling the function $g_k(\cdot)$ as $s_k(\mathbf{x}) = \sqrt{\lambda_k} g_k(\mathbf{x})$, the kernel function can be further presented as

$$K(\mathbf{x}, \mathbf{z}) = \sum_k s_k(\mathbf{x}) \cdot s_k(\mathbf{z}) = [\mathbf{s}(\mathbf{x})]^T \cdot [\mathbf{s}(\mathbf{z})]$$

where $[\mathbf{s}(\mathbf{x})]^T = (s_1(\mathbf{x}), s_2(\mathbf{x}), \ldots, s_l(\mathbf{x}))$ and $[\cdot]^T$ is the transpose operator. Thus, if we further define

$$\nabla \mathbf{s} = \left(\frac{\partial \mathbf{s}(\mathbf{x})}{\partial \mathbf{x}}\right) = \begin{pmatrix} \frac{\partial s_1(\mathbf{x})}{\partial x_1} & \cdots & \frac{\partial s_1(\mathbf{x})}{\partial x_p} \\ \vdots & \vdots & \vdots \\ \frac{\partial s_l(\mathbf{x})}{\partial x_1} & \cdots & \frac{\partial s_l(\mathbf{x})}{\partial x_p} \end{pmatrix}$$

and

$$s_{ij}(\mathbf{x}) = \left(\frac{\partial}{\partial x_i}\mathbf{s}(\mathbf{x})\right)^T \cdot \left(\frac{\partial}{\partial x_j}\mathbf{s}(\mathbf{x})\right)$$

$$= \left(\frac{\partial s_1(\mathbf{x})}{\partial x_i}, \ldots, \frac{\partial s_l(\mathbf{x})}{\partial x_i}\right) \cdot \left(\frac{\partial s_1(\mathbf{x})}{\partial x_j}, \ldots, \frac{\partial s_l(\mathbf{x})}{\partial x_j}\right)^T,$$

as in (8) and (10), it follows that

$$\frac{\partial}{\partial x_i}\frac{\partial}{\partial z_j} K(\mathbf{x}, \mathbf{z})|_{\mathbf{z}=\mathbf{x}} = [\nabla \mathbf{s}(\mathbf{x})]^T \cdot \nabla \mathbf{s}(\mathbf{z}) = \left(\frac{\partial}{\partial x_i}\mathbf{s}(\mathbf{x})\right)^T \cdot \left(\frac{\partial}{\partial x_j}\mathbf{s}(\mathbf{x})\right) = s_{ij}(\mathbf{x}). \ \sharp$$

□

The lemma shows how a mapping **s** is associated with the corresponding kernel function K.

References

1. Liu, X.; He, W. Adaptive kernel scaling support vector machine with application to a prostate cancer image study. *J. Appl. Stat.* **2021**, 1–20. [CrossRef]
2. Crammer, K.; Singer, Y. On the algorithmic implementation of multiclass kernel-based vector machines. *J. Mach. Learn. Res.* **2001**, *2*, 265–292.
3. Maratea, A.; Petrosino, A. Asymmetric Kernel scaling for imbalanced data classification. *Fuzzy Log. Appl.* **2011**, 196–203. [CrossRef]
4. Zhang, Z.; Gao, G.; Shi, Y. Credit risk evaluation using multi-criteria optimization classifier with kernel, fuzzification and penalty factors. *Eur. J. Oper. Res.* **2014**, *237*, 335–348. [CrossRef]
5. Vapnik, V.N.; Vapnik, V. *Statistical Learning Theory*; Wiley: New York, NY, USA, 1998; Volume 1.
6. Menardi, G.; Torelli, N. Training and assessing classification rules with imbalanced data. *Data Min. Knowl. Discov.* **2014**, *28*, 92–122. [CrossRef]

7. Kreßel, U.H.G. Pairwise classification and support vector machines. In *Advances in Kernel Methods*; MIT Press: Cambridge, MA, USA, 1999; pp. 255–268.
8. Suykens, J.A.; Vandewalle, J. Least squares support vector machine classifiers. *Neural Process. Lett.* 1999, 9, 293–300. [CrossRef]
9. Suykens, J.A.; Vandewalle, J. Multiclass least squares support vector machines. In Proceedings of the International Joint Conference on Neural Networks, IJCNN'99, Washington, DC, USA, 10–16 July 1999; Volume 2, pp. 900–903.
10. Xia, X.L.C.; Li, K. A sparse multi-class least-squares support vector machine. In Proceedings of the IEEE International Symposium on Industrial Electronics, Cambridge, UK, 30 June–2 July 2008, pp. 1230–1235.
11. Fung, G.M.; Mangasarian, O.L. Multicategory proximal support vector machine classifiers. *Mach. Learn.* 2005, 59, 77–97. [CrossRef]
12. Fung, G.M.; Mangasarian, O. Proximal support vector machine classifiers. *Mach. Learn.* 2002, 1, 21.
13. Zhang, Y.; Fu, P.; Liu, W.; Chen, G. Imbalanced data classification based on scaling kernel-based support vector machine. *Neural Comput. Appl.* 2014, 25, 927–935. [CrossRef]
14. He, X.; Wang, Z.; Jin, C.; Zheng, Y.; Xue, X. A simplified multi-class support vector machine with reduced dual optimization. *Pattern Recognit. Lett.* 2012, 33, 71–82. [CrossRef]
15. Mazurowski, M.A.; Habas, P.A.; Zurada, J.M.; Lo, J.Y.; Baker, J.A.; Tourassi, G.D. Training neural network classifiers for medical decision making: The effects of imbalanced datasets on classification performance. *Neural Netw.* 2008, 21, 427–436. [CrossRef] [PubMed]
16. Chawla, N.; Japkowicz, N.; Kolcz, A. *Special Issue on Learning from Imbalanced Datasets, Sigkdd Explorations*; ACM SIGKDD: New York, NY, USA, 2004; Volume 6, pp. 1–6
17. Daskalaki, S.; Kopanas, I.; Avouris, N. Evaluation of classifiers for an uneven class distribution problem. *Appl. Artif. Intell.* 2006, 20, 381–417. [CrossRef]
18. Chawla, N.V.; Japkowicz, N.; Kotcz, A. Editorial: special issue on learning from imbalanced data sets. *ACM Sigkdd Explor. Newsl.* 2004, 6, 1–6. [CrossRef]
19. Tang, Y.; Zhang, Y.Q.; Chawla, N.V.; Krasser, S. SVMs modeling for highly imbalanced classification. *Syst. Man Cybern. Part B Cybern. IEEE Trans.* 2009, 39, 281–288. [CrossRef] [PubMed]
20. Wang, L.; Shen, X. On L1-norm multiclass support vector machines. *J. Am. Stat. Assoc.* 2007, 102, 583–594.
21. Friedman, J.; Hastie, T.; Tibshirani, R. *The Elements of Statistical Learning*; Springer Series in Statistics; Springer: Berlin/Heidelberg, Germany, 2001; Volume 1.
22. Wu, S.; Amari, S.I. Conformal transformation of kernel functions: A data-dependent way to improve support vector machine classifiers. *Neural Process Lett.* 2002, 15, 59–67. [CrossRef]
23. Williams, P.; Li, S.; Feng, J.; Wu, S. Scaling the kernel function to improve performance of the support vector machine. In *Advances in Neural Networks–ISNN 2005*; Springer: Berlin/Heidelberg, Germany, 2005; pp. 831–836.
24. Maratea, A.; Petrosino, A.; Manzo, M. Adjusted F-measure and kernel scaling for imbalanced data learning. *Inf. Sci.* 2014, 257, 331–341. [CrossRef]
25. Fawcett, T. ROC graphs: Notes and practical considerations for researchers. *Mach. Learn.* 2004, 31, 1–38.

Article

Damped Newton Stochastic Gradient Descent Method for Neural Networks Training

Jingcheng Zhou [1,2], Wei Wei [1,2,3,4,*], Ruizhi Zhang [1,2] and Zhiming Zheng [1,2,3,4]

1. School of Mathematical Sciences, Beihang University, Beijing 100191, China; 17374347@buaa.edu.cn (J.Z.); ruizhiz@buaa.edu.cn (R.Z.); zzheng@pku.edu.cn (Z.Z.)
2. Key Laboratory of Mathematics, Informatics and Behavioral Semantics, Ministry of Education, Beijing 100191, China
3. Peng Cheng Laboratory, Shenzhen 518055, China
4. Beijing Advanced Innovation Center for Big Data and Brain Computing, Beihang University, Beijing 100191, China
* Correspondence: weiw@buaa.edu.cn

Abstract: First-order methods such as stochastic gradient descent (SGD) have recently become popular optimization methods to train deep neural networks (DNNs) for good generalization; however, they need a long training time. Second-order methods which can lower the training time are scarcely used on account of their overpriced computing cost to obtain the second-order information. Thus, many works have approximated the Hessian matrix to cut the cost of computing while the approximate Hessian matrix has large deviation. In this paper, we explore the convexity of the Hessian matrix of partial parameters and propose the damped Newton stochastic gradient descent (DN-SGD) method and stochastic gradient descent damped Newton (SGD-DN) method to train DNNs for regression problems with mean square error (MSE) and classification problems with cross-entropy loss (CEL). In contrast to other second-order methods for estimating the Hessian matrix of all parameters, our methods only accurately compute a small part of the parameters, which greatly reduces the computational cost and makes the convergence of the learning process much faster and more accurate than SGD and Adagrad. Several numerical experiments on real datasets were performed to verify the effectiveness of our methods for regression and classification problems.

Keywords: stochastic gradient descent; damped Newton; convexity

1. Introduction

First-order methods are popularly used to train deep neural networks (DNNs), such as stochastic gradient descent (SGD) [1] and its variants which use momentum and acceleration [2] and an adaptive learning rate [3]. At each iteration, SGD calculates the gradient on only a small batch instead of the whole training data. Such randomness introduced by sampling the small batch can lead to the better generalization of the DNNs [4]. Recently, there is a lot of work attempting to come up with more efficient first-order methods beyond SGD [5]. First-order methods are easy to implement and only require moderate computational cost per iteration; however, the requirement of adjusting their hyper-parameters (such as learning rate) becomes a complex task, and they are often slow to flee from areas when the loss function's Hessian matrix is ill-conditioned [6].

Second-order stochastic methods were also proposed for training DNNs because they take far fewer iterations. Second-order stochastic methods especially have the ability to escape from regions where the Hessian matrix of the loss function is ill-conditioned and provide adaptive learning rates. However, the main problem here is that it is practically impossible to compute and invert a full Hessian matrix due to the massive parameters of DNNs and the Hessian matrix is not always positive definite [7]. Efforts to conquer this problem include Kronecker-factored approximate [8–10], Hessian-free inexact Newton methods [11–14], stochastic L-BFGS methods, Gauss–Newton [15] and natural gradient methods [16–18] and diagonal scaling methods [19] which cut the cost by approximating the Hessian matrix. In addition, at the base of these methods, more algorithms are generated, such as SMW-GN and SMW-NG, which use the Sherman–Morrison–Woodbury formula to cut the cost of computing the inverse matrix [20] and GGN that combining GN and NTK [21]. There are also some variances of the Quasi–Newton methods for approximating the Hessian matrix [22–25]. However, most of these second-order methods lose too much information of the Hessian matrix. They lose a part of information by obtaining the approximate Hessian matrix, and further loss by adding regular terms to make the Hessian matrix positive definite.

1.1. Our Contributions

Our main contribution is the development of methods to train DNNs that combine first-order information and some second-order information, while keeping the computational cost of each iteration comparable to that required by first-order methods. To achieve this, we propose new damped Newton stochastic gradient descent methods that can be efficiently implemented. Our methods adjust the iteration of SGD by changing the iteration of the parameters of the last layer of DNNs. We used the damped Newton method for the iteration of the parameters of the last layer of DNNs, which makes the loss function decrease more rapidly. In addition, a quicker descent can exceed the cost of computing the Hessian matrix of the parameters of the layer. We elaborated the availability of our methods with numerical experiments, and compared these with those of the first-order method (SGD) and Adagrad.

The novelty of our work is that we computed the Hessian matrices of the last layer parameters and the penultimate layer of the loss function by mathematical formula and mathematically and rigorously demonstrated that the Hessian matrices are positive semi-definite. Furthermore, we then propose the DN-SGD and SGD-DN algorithms, which let the parameters of the last layer iterate with the variational damped Newton method.

To the best of our knowledge, we were the first to directly calculate the loss function of the last layer and the penultimate layer parameters of the Hessian matrix with mathematical expressions, and demonstrate the property of the semi-positive definite of the Hessian matrix, and then use the property of the semi-positive definite to put forward our algorithms. Most other works have sought to approximate the whole Hessian matrix of all parameters by various methods, and then make it positive definite by adding a quantitative matrix.

1.2. Related Work

SMW-GN and SMW-NG [20] adjust the Gauss–Newton method (GN) and natural gradient method (NG) which use the Sherman–Morrison–Woodbury (SMW) formula to compute the inverse matrix to cut the cost of computing second-order information. Goldfarb [22] approximates the Hessian matrix by a block-diagonal matrix and uses the structure of the gradient and Hessian to further approximate these blocks, each of which corresponds to a layer, as the Kronecker product of two much smaller matrices. KF-QN-CNN [23] approximates the Hessian matrix by a layer-wise block diagonal matrix and each layer's diagonal block is further approximated by a Kronecker product corresponding to the structure of the Hessian restricted to that layer.

The Adagrad algorithm [3] individually adapts to the learning rate of all model parameters by scaling all model parameters inversely to the square root of the sum of their historical square values. The learning rate of the parameter with the largest loss partial derivative decreases rapidly, while the learning rate of the parameter with the smallest partial derivative decreases relatively little. The net effect is a greater advance in the flat direction of the parameter space [26].

2. Main Results

2.1. Feed-Forward Neural Networks

Consider a simple fully connected neural network with $L+1$ layers. For each layer, the number of nodes is n_l, $0 \le l \le L$. In the following expressions, we put the bias into the weights for simplicity. Given an input x, add a component for x to make the input as $x^{(0)} = (1, x^T)^T$ for the first layer. For the l-th layer, add a component for $x^{(l-1)}$, so $x^{(l)} = \sigma^{(l)}(W^{(l)} x^{(l-1)})$, where $W^{(l)} = (w_1^{(l)}, w_2^{(l)}, \ldots, w_{n_l}^{(l)})^T \in R^{n_l \times (n_{l-1}+1)}$, $w_j^{(l)} = (w_{0j}^{(l)}, w_{1j}^{(l)}, \ldots, w_{n_{l-1}j}^{(l)})^T \in R^{n_{l-1}+1}$ which denote the parameters between the $(l-1)$-th layer and the j-th node of l-th layer, $\sigma^{(l)} : R \to R$ is a nonlinear activation function and when $\sigma^{(l)}$ is used on a vector, which means $\sigma^{(l)}$ operates on every component of the vector, and the output $f(\theta, x) = \sigma^{(L)}(W^{(L)} x^{(L-1)})$ where $\theta = \left(\text{vec}((W^{(1)})^T), \text{vec}((W^{(2)})^T), \ldots, \text{vec}((W^{(L)})^T) \right)^T$ in which vec(W) vectorizes the matrix W by concatenating its columns.

Given the training data $(X, Y) = \{(x_1, y_1), (x_2, y_2), \ldots, (x_m, y_m)\}$, for training the neural networks, the loss function for minimization is generally defined as

$$L(\theta) = \frac{1}{m} \sum_{i=1}^{m} L_i(\theta, X, Y) = \frac{1}{m} \sum_{i=1}^{m} l(f(\theta, x_i), y_i), \tag{1}$$

where $l(f(\theta, x_i), y_i)$ is a loss function. For the learning tasks, the standard mean square error (MSE) is always used for the regression problems and the cross-entropy loss (CEL) is used for the classification problems. Note $\theta \in R^n$ where $n = \sum_{i=1}^{L} n_l(n_{l-1}+1)$.

2.2. Convexity of Partial Parameters of the Loss Function

Theorem 1. *Given a neural network such as the one above using MSE for a loss function of regression problems, then the number of the nodes for the L-th layer is one. In the last layer, the activation function is used as identical function for regression problems. Then, the Hessian matrix of $L(\theta)$ to the parameters of last layer is positive semi-definite. When $\sigma^{(L-1)}(x) = relu(x) = max(0, x)$, the Hessian matrix of $L(\theta)$ to the parameters of last but one layer is also positive semi-definite.*

Proof of Theorem 1. For simplicity, consider one single input first. Given the input x_i, the loss function should be:

$$L(W^{(L)}) = \left(W^{(L)} x_i^{(L-1)} - y_i \right)^2$$

and its gradient vector as well as Hessian matrix should be:

$$\frac{\partial L(W^{(L)})}{\partial W^{(L)}} = 2 \left(W^{(L)} x_i^{(L-1)} - y_i \right) x_i^{(L-1)}, \tag{2}$$

$$\frac{\partial^2 L(W^{(L)})}{\partial (W^{(L)})^2} = 2 x_i^{(L-1)} (x_i^{(L-1)})^T = A. \tag{3}$$

With the knowledge of matrix [27], if $x_i^{(L-1)}$ is not a null vector, it can easily be obtained that A is a positive semi-definite matrix with rank one. As for multiple inputs, the Hessian is the addition of several positive semi-definite matrices, so the Hessian matrix is also positive semi-definite.

If the activation function $\sigma^{(L-1)}(x) = max(0, x)$, we can obtained the loss function:

$$L(W^{(L-1)}) = (W^{(L)} \sigma^{(L-1)} (W^{(L-1)} x_i^{(L-2)}) - y_i)^2$$

and the gradient vector as well as the Hessian matrix:

$$\frac{\partial L}{\partial w_j^{(L-1)}} = 2(W^{(L)}\sigma^{(L-1)}(W^{(L-1)}x_i^{(L-2)}) - y_i)W_j^{(L)}\sigma'((w_j^{(L-1)})^T x_i^{(L-2)})x_i^{(L-2)}, \quad (4)$$

$$\frac{\partial^2 L}{\partial w_j^{(L-1)} \partial w_k^{(L-1)}} = 2W_j^{(L)} W_k^{(L)} \sigma'((w_j^{(L-1)})^T x_i^{(L-2)})\sigma'((w_k^{(L-1)})^T x_i^{(L-2)}) x_i^{(L-2)} (x_i^{(L-2)})^T, \quad (5)$$

$$\frac{\partial^2 L}{\partial (v^{L-1})^2} = B \otimes C, \quad (6)$$

where $\otimes$ denotes Kronecker product and:

$$v^{(L-1)} = ((w_1^{(L-1)})^T, (w_2^{(L-1)})^T, \cdots, (w_{n_{L-1}}^{(L-1)})^T)^T,$$

$$B \in R^{n_{L-1} \times n_{L-1}}, B_{jk} = 2W_j^{(L)} W_k^{(L)} \sigma'((w_j^{(L-1)})^T x_i^{(L-2)}) \sigma'((w_k^{(L-1)})^T x_i^{(L-2)}),$$

$$C = x_i^{(L-2)} (x_i^{(L-2)})^T.$$

Let $u_j = W_j^{(L)} \sigma'((w_k^{(L-1)})^T x_i^{(L-2)})$, $u = (u_1, u_2, \cdots, u_{n_{L-1}})^T$, we can obtain $B = 2uu^T$. Thus, it is easy to see that B and C are positive semi-definite matrix. Since the Kronecker product of positive semi-definite matrices is also positive semi-definite [28], the Hessian matrix $B \otimes C$ is positive semi-definite. Hence, the Hessian of the multiple inputs is also positive semi-definite. □

Theorem 2. *We take a neural network such as the one above using CEL for loss function of classification problems. In the last layer, softmax is used as an activation function for classification problems. Then, the Hessian matrix of $L(\theta)$ to the parameters of last layer is positive semi-definite.*

Proof of Theorem 2. For simplicity, consider one single input first. Given the input x_i, the loss function should be:

$$L = -\sum_{j=1}^{n_L} Y_j \ln x_{ij}^{(L)},$$

where Y_j is 1 when x_i is the j-th class, otherwise 0. $x_{ij}^{(L)} = \dfrac{e^{(w_j^{(L)})^T x_i^{(L-1)}}}{\sum_{j=1}^{n_L} e^{(w_j^{(L)})^T x_i^{(L-1)}}}$, so

$$L = -\sum_{j=1}^{n_L} Y_j (w_j^{(L)})^T x_i^{(L-1)} + \ln(D),$$

where $D = \sum_{j=1}^{n_L} e^{(w_j^{(L)})^T x_i^{(L-1)}}$.

Then, compute the gradient vector as well as Hessian matrix:

$$\frac{\partial L}{\partial w_j^{(L)}} = \left(\frac{e^{(w_j^{(L)})^T x_i^{(L-1)}}}{D} - Y_j \right) x_i^{(L-1)}, \quad (7)$$

$$\frac{\partial^2 L}{\partial w_j^{(L)} \partial w_k^{(L)}} = \begin{cases} \dfrac{e^{(w_j^{(L)})^T x_i^{(L-1)}} (D - e^{(w_j^{(L)})^T x_i^{(L-1)}})}{D^2} x_i^{(L-1)} (x_i^{(L-1)})^T, j = k, \\ -\dfrac{e^{(w_j^{(L)})^T x_i^{(L-1)}} e^{(w_k^{(L)})^T x_i^{(L-1)}}}{D^2} x_i^{(L-1)} (x_i^{(L-1)})^T, j \neq k. \end{cases} \quad (8)$$

Let $v^{(L)} = ((w_1^{(L)})^T, (w_2^{(L)})^T, \ldots, (w_{n_L}^{(L)})^T)^T$, then we have:

$$\frac{\partial^2 L}{\partial (v^{(L)})^2} = E \otimes F, \tag{9}$$

where:

$$F = \frac{1}{D^2} x_i^{(L-1)} (x_i^{(L-1)})^T,$$

$$E \in R^{n_L \times n_L}, E_{jk} = \begin{cases} e^{(w_j^{(L)})^T x_i^{(L-1)}} (D - e^{(w_j^{(L)})^T x_i^{(L-1)}}), j = k, \\ -e^{(w_j^{(L)})^T x_i^{(L-1)}} e^{(w_k^{(L)})^T x_i^{(L-1)}}, j \neq k. \end{cases} \tag{10}$$

For any vector $z = (z_1, z_2, \cdots, z_{n_L})^T \in R^{n_L}$:

$$z^T E z = \sum_{jk}^{n_L} z_j z_k E_{jk} = \frac{1}{2} \sum_{jk}^{n_L} e^{(w_j^{(L)})^T x_i^{(L-1)}} e^{(w_k^{(L)})^T x_i^{(L-1)}} (z_j - z_k)^2 \geq 0. \tag{11}$$

Thus, E and F are positive semi-definite matrices. Because the Kronecker product of positive semi-definite matrices is also positive semi-definite [28], the Hessian matrix $E \otimes F$ is positive semi-definite and the Hessian matrix of the condition of multiple inputs is also positive semi-definite. □

3. Our Innovation: Damped Newton Stochastic Gradient Descent

As for the property of the loss function, the convexity of the loss function can be determined for the last layer of DNNs. Thus, we can use the damped Newton method for the parameters of the last layer and SGD for other layers. For each iteration, if the damped Newton method is used before SGD, it is called damped Newton stochastic gradient descent (DN-SGD). Otherwise, if SGD is used before the damped Newton method, wit is called stochastic gradient descent damped Newton method (SGD-DN).

The difficulty of neural network optimization is that the loss function is a non-convex function and there is no good method for the optimization of a non-convex function. One of the methods commonly used at present is the stochastic gradient descent method, however, the speed of the gradient descent method is slow. Our method improves the stochastic gradient descent method and accelerates its convergence speed.

Our method combines the advantages of the stochastic gradient descent and damped Newton. On the one hand, the main part of the parameters of DNNs using the SGD method for training guarantees that the computing cost changes little. On the other hand, the parameters of the last layer using DN-SGD or SGD-DN for training can make the learning process converge more quickly with just a little more computing cost. When the training data are numerous, a lesser number of iterations can significantly cut time for training the DNNs and overcome the cost of computing the second-order information of the last layer.

3.1. Defect of Methods Approximating the Hessian Matrix

The quasi-Newton method, similarly to others, approximates the Hessian matrix to cut the computing cost, however, all these methods consider the whole Hessian matrix and lose significant information to approximate. In addition, they add the scalar matrix to make the Hessian matrix positive definite.

In traditional damped Newton method, adding scalar matrix is also used to make the Hessian matrix positive definite. The iteration is

$$\theta^{(t+1)} = \theta^{(t)} - (H^{(t)} + \lambda I)^{-1} g^{(t)}.$$

This method is hard to put into practice for the uncertainty of λ. We cannot know the eigenvalues of the Hessian matrix so that we cannot precisely set λ. If λ is set too big, the

Hessian matrix is close to a scalar matri; however, if λ is too small, the Hessian matrix is not sufficient to be positive definite.

3.2. Set λ Precisely

Our optimization method was based on a more elaborated analysis for the Hessian matrix, in which the Hessian matrix was not entirely used for accelerating the training. We only used the last layer's Hessian matrix by damped Newton method. The iteration was

$$\theta^{(t+1)} = \theta^{(t)} - (H^{(t)} + \lambda I)^{-1} g^{(t)},$$

where $\lambda = \beta H_{max}^{(t)} + \alpha$ in which $H_{max}^{(t)}$ is the maximum value of the element of the $H^{(t)}$ and α is the damping coefficient. $H_{max}^{(t)}$ is to make the matrix positive definite and lower the conditional number at the same time, and β is to more precisely adjust the conditional number. α is only to prevent the condition that $H^{(t)}$ is too bad to be a null matrix or very close to a null matrix, such as the results obtained for the dataset of Shot Selection.

3.3. Last Layer Makes Front Layers Converge Better

We use the chain rule to obtain:

$$\delta^{(l)} = (w^{(l+1)T} \delta^{(l+1)}) \odot \sigma'(x^{(l)}),$$

where $\delta^{(l)} = \dfrac{\partial L}{\partial x^{(l)}}$. We use the damped Newton method to train the last layer so that they can quickly obtain the stationary point. When the parameters of the last layer almost obtain the stationary point, the gradient almost turns zero. Thus, the gradient of the parameters of front layers are almost zero, so that they also almost the stationary point.

4. Algorithm

In this section, the DN-SGD and SGD-DN methods are summarized in Algorithms 1 and 2.

Algorithm 1: DN-SGD.

Input: Training Set $S_M = \{(x_1, y_1), (x_2, y_2), \ldots, (x_M, y_M)\}$, hyper-parameter β, α, learning rate l, batchsize N

1 Initialize the network parameter $\theta^{(0)}, t = 0$
2 **for** $t = 0, 1, 2, \ldots$ **do**
3   Randomly select a mini-batch $S_N \subset S_M$ of size N
4   Compute $H_{last}^{(t)}, g_{last}^{(t)}$ for the last layer and $H_{max}^{(t)} = \max(H_{last}^{(t)})$
5   Compute $p^{(t)} = (H_{last}^{(t)} + \lambda I)^{-1}, \lambda = \beta H_{max}^{(t)} + \alpha$
6   Set $\theta_{last}^{(t+1)} = \theta_{last}^{(t+1)} - p^{(t)} g_{last}^{(t)}$
7   Compute $g_{front}^{(t)}$ for other layers
8   Set $\theta_{front}^{(t+1)} = \theta_{front}^{(t)} - l g_{front}^{(t)}$
9 **end for**

The only difference between the DN-SGD (Algorithm 1) and SGD-DN (Algorithm 2) methods is that in each iteration, the damped Newton method or the stochastic gradient descent is used first.

Algorithm 2: SGD-DN.

Input: Training Set $S_M = \{(x_1, y_1), (x_2, y_2), \ldots, (x_M, y_M)\}$, hyper-parameter β, α, learning rate l, batchsize N

1 Initialize the network parameter $\theta^{(0)}, t = 0$
2 **for** $t = 0, 1, 2, \ldots$ **do**
3   Randomly select a mini-batch $S_N \subset S_M$ of size N
4   Compute $g_{front}^{(t)}$ for other layers
5   Set $\theta_{front}^{(t+1)} = \theta_{front}^{(t)} - l g_{front}^{(t)}$
6   Compute $H_{last}^{(t)}, g_{last}^{(t)}$ for the last layer and $H_{max}^{(t)} = \max(H_{last}^{(t)})$
7   Computer $p^{(t)} = (H_{last}^{(t)} + \lambda I)^{-1}, \lambda = \beta H_{max}^{(t)} + \alpha$
8   Set $\theta_{last}^{(t+1)} = \theta_{last}^{(t+1)} - p^{(t)} g_{last}^{(t)}$
9 **end for**

5. Numerical Example

5.1. Regression Problem

The performance of the algorithms DN-SGD and SGD-DN was compared with SGD and Adagrad, and the experiments were performed on several regression problems, and both training loss and testing loss are provided. The datasets are scaled to have normal distribution.

House Prices: The training set is of size 1361, and the test set is of size 100. A neural network with one hidden layer of size 5 and sigmoid activation function was used, i.e., $(n_0, n_1, n_2) = (288, 5, 1)$, where the first and last layers are the size of the input and output, respectively. The output layer has no activation function except with MSE, and the learning rate for SGD was set to be 0.01—the same as other methods for comparison purposes. Each algorithm was run for 20 epochs. The website is https://www.kaggle.com/c/house-prices-advanced-regression-techniques/data (accessed on 25 March 2021). The results are presented in Figure 1.

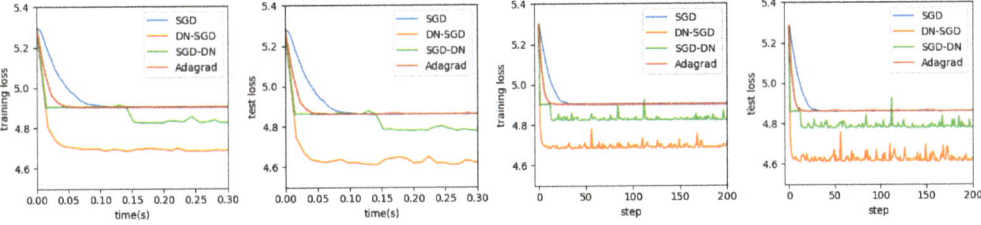

Figure 1. Results on House Price regression problem with $N = 50$, $\beta = 0.1$, $\alpha = 0$, $l = 0.01$.

Housing Market: The training set is of size 20,473, and the test set is of size 10,000. A neural network with one hidden layer of size 4 and sigmoid activation function was used, i.e., $(n_0, n_1, n_2) = (451, 4, 1)$, where the first and last layers are the size of the input and output, respectively. The output layer has no activation function except with MSE, and the learning rate for SGD was set to be 0.001—the same as other methods for comparison purposes. Each algorithm was run for five epochs. The website is https://www.kaggle.com/c/sberbank-russian-housing-market/data (accessed on 25 March 2021). The results are presented in Figure 2.

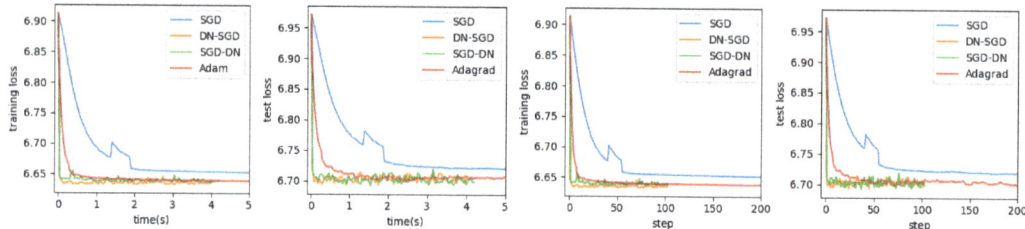

Figure 2. Results on Housing Market regression problem with $N = 200$, $\beta = 0.1$, $\alpha = 0$, $l = 0.01$.

Cabbage Price: The training set is of size 2423, and the test set is of size 500. A neural network with one hidden layer of size 6 and sigmoid activation function was used, i.e., $(n_0, n_1, n_2) = (4, 6, 1)$, where the first and last layers are the size of the input and output, respectively. The output layer has no activation function except with MSE, and the learning rate for SGD was set to be 0.01—the same as other methods for comparison purposes. Each algorithm was run for five epochs. The website is https://www.kaggle.com/c/regression-cabbage-price/data (accessed on 25 March 2021). The results are presented in Figure 3.

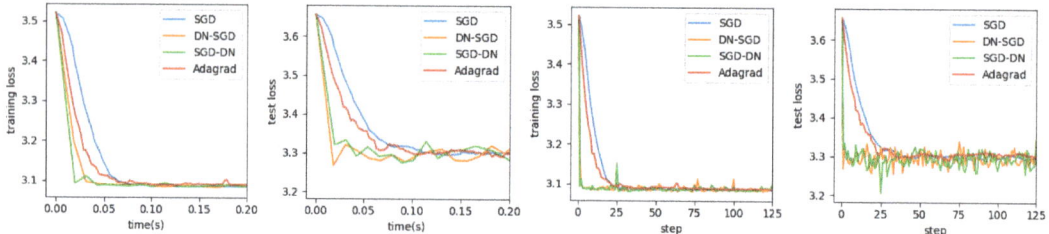

Figure 3. Results on Cabbage Price regression problem with $N = 100$, $\beta = 0.1$, $\alpha = 0$, $l = 0.01$.

5.2. Classification Problem

The performance of the algorithms DN-SGD and SGD-DN was compared with SGD Adagrad, the experiments were performed on several classification problems, and both training loss and testing loss are provided. The datasets were scaled to have normal distribution.

Shot Selection: The training set is of size 20,698, and the test set is of size 10,000. A neural network with one hidden layer of size 6 and sigmoid activation function was used, i.e., $(n_0, n_1, n_2) = (228, 6, 2)$, where the first and last layers are the size of the input and output, respectively. The output layer has softmax with cross entropy, and the learning rate for SGD was set to be 0.01—the same as other methods for comparison purposes. Each algorithm was run for 25 epochs. The website is https://www.kaggle.com/c/kobe-bryant-shot-selection/data (accessed on 25 March 2021). The results are presented in Figure 4.

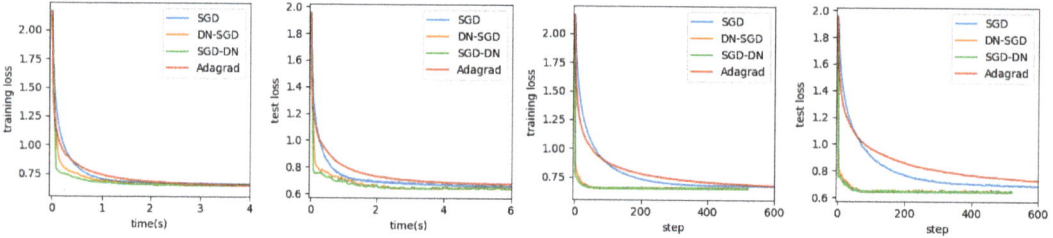

Figure 4. Results on Shot Selection classification problem with $N = 200$, $\beta = 1$, $\alpha = 0.01$, $l = 0.01$.

Categorical Feature Encoding: The training set is of size 30,000, and the test set is of size 10,000. A neural network with one hidden layer of size 6 and sigmoid activation function was used, i.e., $(n_0, n_1, n_2) = (96, 4, 2)$, where the first and last layers are the size of the input and output, respectively. The output layer has softmax with cross entropy, and the learning rate for SGD was set to be 0.01—the same as other methods for comparison purposes. Each algorithm was run for 20 epochs. The website is https://www.kaggle.com/c/cat-in-the-dat/data (accessed on 25 March 2021). The results are presented in Figure 5.

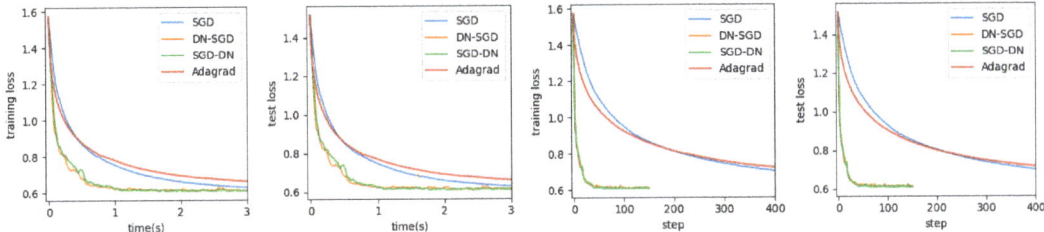

Figure 5. Results on Categorical Feature Encoding classification problem with $N = 200$, $\beta = 1$, $\alpha = 0$, $l = 0.01$.

Santander Customer Satisfaction: The training set is of size 30,000, and the test set is of size 10,000. A neural network with one hidden layer of size 4 and sigmoid activation function was used, i.e., $(n_0, n_1, n_2) = (369, 4, 2)$, where the first and last layers are the size of the input and output, respectively. The output layer has softmax with cross entropy, and the learning rate for SGD was set to be 0.01—the same as other methods for comparison purposes. Each algorithm was run for 20 epochs. The website is https://www.kaggle.com/c/santander-customer-satisfaction/data (accessed on 25 March 2021). The results are presented in Figure 6.

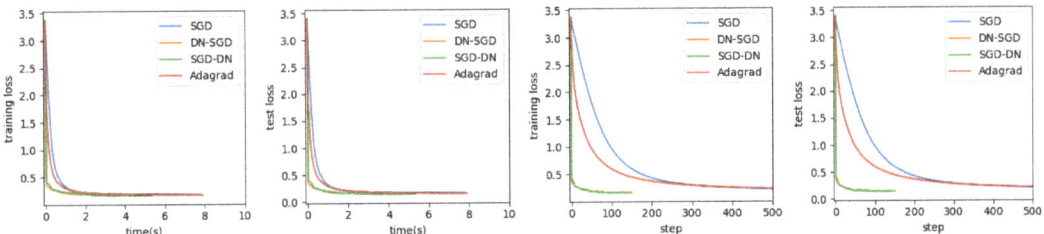

Figure 6. Results on Santander Customer Satisfaction classification problem with $N = 200$, $\beta = 1$, $\alpha = 0$, $l = 0.01$.

6. Discussion of Results

From the experimental results, it can be seen that DN-SGD and SGD-DN are always faster than SGD and Adagrad in terms of both steps and time, which is consistent with the provided analysis. In addition, when the dataset is huge, the DN-SGD and SGD-DN can be more efficient because the advantage of quick descent can exceed the cost of computing the second-order information. DN-SGD is to find the optimal point for the parameters of last layer and then update other parameters using SGD in each iteration. SGD-DN is to update parameters, except the last layer using SGD, and then find the optimal point for the parameters of the last layer in each iteration. They both make full use of the convergence and rate of convergence of convexity of the last layer. We will analyze the experimental results in detail below.

6.1. Regression Problem

We performed our experiments on three real datasets: House Price, Housing Market, and Cabbage Price. In every experiment, we maintained the same DNNs model and initialization parameters to train DNNs using SGD, Adagrad, DN-SGD and SGD-DN. In

Table 1, we can see that DN-SGD and SGD-DN perform consistently and they both perform better than SGD and Adagrad in terms of time and steps in both training datasets and testing datasets. The training loss curves and test loss curves of different optimization algorithms for regression tasks are shown in Figures 1–3. Our proposed methods converge much faster than SGD and Adagrad. We can see from Figure 1 that the loss of using our DN-SGD method quickly decreases to a minimum point in no more than ten steps in both training loss and test loss in all three datasets. In the dataset House Price, on the contrary, using SGD, the loss decays much slower than our method in terms of clock time and steps and the minimum point is higher than DN-SGD. At the same time, Adagrad takes about the same amount of time to arrive a higher minimum point than DN-SGD. SGD-DN takes more time but also has a lower minimum point than Adagrad. Thus, DN-SGD and SGD-DN have a better generalization than SGD and Adagrad. In the dataset Housing Market, DN-SGD is nearly 20 times faster in converging to the stationary point than SGD and Adagrad in terms of time and steps in both training loss and test loss. In the dataset Cabbage Price, DN-SGD and SGD-DN also perform better than SGD and Adagrad in almost half the time and one-sixth of the steps. All of this suggests that SGD-DN and DN-SGD have a better convergence than SGD and Adagrad.

Table 1. Running time, iteration steps and minimum loss for regression problem. The numbers in bold are the optimal values.

	Algorithm	House Prices			Housing Market			Cabbage Price		
		Time (s)	Steps	Loss	Time (s)	Steps	Loss	Time (s)	Steps	Loss
Training	SGD	0.15	49	4.91	3.09	104	6.66	0.12	32	3.11
	DN-SGD	0.06	5	4.69	0.15	5	6.63	0.05	5	3.10
	SGD-DN	0.15	22	4.81	0.73	39	6.64	0.048	5	3.10
	Adagrad	0.05	23	4.91	2.97	86	6.64	0.10	48	3.10
Test	SGD	0.12	51	4.86	3.91	147	6.72	0.14	46	3.30
	DN-SGD	0.05	5	4.62	0.17	6	6.70	0.04	7	3.28
	SGD-DN	0.15	23	4.78	0.96	42	6.70	0.10	25	3.21
	Adagrad	0.05	23	4.86	3.23	75	6.70	0.11	41	3.30

6.2. Classification Problem

For the classification task, we also experimented on three real datasets: Shot Selection, Categorical Feature Encoding, and Santander Customer Satisfaction. In every experiment, we also maintained the same DNNs model and initialization parameters to train DNNs using SGD, Adagrad, DN-SGD and SGD-DN. Obviously, we can see from Table 2 that SGD-DN and DN-SGD perform better than SGD and Adagrad, however, the performance is worse than for regression task. This is because the Hessian matrix for the classification task has a bigger size than the regression task so computing the inverse matrix takes more time. In addition, we can see that DN-SGD and SGD-DN perform consistently in both time and steps. In the datasets Shot Selection and Categorical Feature Encoding, the loss of using our DN-SGD method quickly decreases to a minimum point in several steps. On the contrary, using SGD and Adagrad, the loss decays much slower than our method in terms of clock time and steps and the minimum point is higher than DN-SGD and SGD-DN. Thus, DN-SGD and SGD-DN have a better generalization than SGD and Adagrad. In the dataset Santander Customer Satisfaction, DN-SGD, SGD-DN, SGD and Adagrad take almost the same time to reach the same stationary point, however, DN-SGD and SGD-DN take less steps than SGD and Adagrad. Anyway, SGD-DN and DN-SGD perform better in convergence and generalization. In Table 2, what we have in bold is the optimal value.

Table 2. Running time, iteration steps and minimum loss for classification problem. The numbers in bold are the optimal values.

	Algorithm	Shot Selection			Categorical Feature			Santander Customer		
		Time (s)	Steps	Loss	Time (s)	Steps	Loss	Time (s)	Steps	Loss
Training	SGD	2.89	476	0.70	2.76	451	0.64	1.98	451	0.23
	DN-SGD	1.49	39	**0.69**	**1.13**	49	0.62	1.93	73	0.22
	SGD-DN	**1.43**	37	0.69	1.13	49	0.62	1.93	74	0.22
	Adagrad	3.11	531	0.70	2.73	432	0.67	1.95	446	0.23
Test	SGD	4.13	425	0.71	2.74	447	0.64	1.97	448	0.23
	DN-SGD	2.19	51	0.64	1.09	49	0.62	1.92	71	0.22
	SGD-DN	**2.15**	49	0.64	1.09	49	0.62	1.92	73	0.22
	Adagrad	4.27	437	0.73	2.71	425	0.67	1.94	439	0.23

7. Conclusions and Future Research Directions

In this paper, we propose the DN-SGD and SGD-DN methods combining the stochastic gradient descent and damped Newton method, which perform better than SGD and Adagrad in several datasets for training neural networks. These methods also have the potential for application on more sophisticated models such as CNNs, ResNet and even GNNs which are proposed on the basis of multilayer neural network (MLP). These complicated models can have convexity for the parameters of the last layer with a high probability, therefore we can make full use of the convexity to come up with the corresponding efficient algorithms. However, it is very hard to do this mathematically due to the complexity of computing the Hessian matrix in mathematical symbols and the great difficulty of using the naked eye to determine whether the Hessian matrix is positive definite or semi-positive definite. Nonetheless, we think it is a challenging and attracting topic in the area of deep learning optimization. Another promising future research topic is the study of the convexity of the Hessian matrix for each hidden layer, which is more difficult to do in rigorous mathematical methods.

Author Contributions: Conceptualization, J.Z., R.Z. and W.W.; methodology, J.Z.; software, J.Z.; validation, J.Z., W.W. and Z.Z.; formal analysis, J.Z.; investigation, J.Z.; resources, J.Z.; data curation, J.Z.; writing—original draft preparation, J.Z., R.Z.; writing—review and editing, J.Z., W.W.; visualization, J.Z.; supervision, W.W. and Z.Z.; project administration, J.Z. All authors have read and agreed to the published version of the manuscript.

Funding: This work is supported by the Fundamental Research Funds for the Central Universities, the Research and Development Program of China (No.2018AAA0101100), the Beijing Natural Science Foundation (1192012, Z180005) and National Natural Science Foundation of China (No.62050132).

Institutional Review Board Statement: Not applicable.

Informed Consent Statement: Not applicable.

Data Availability Statement: Not applicable.

Acknowledgments: We thank Zhenyu Shi for their detailed guidance on the paper layout.

Conflicts of Interest: The authors declare no conflict of interest.

References

1. Gardner, W.A. Learning characteristics of stochastic-gradient-descent algorithms: A general study, analysis, and critique. *Signal Process.* **1984**, *6*, 113–133. [CrossRef]
2. Qian, N. On the momentum term in gradient descent learning algorithms. *Neural Netw.* **1999**, *12*, 145–151. [CrossRef]
3. Duchi, J.; Hazan, E.; Singer, Y. Adaptive subgradient methods for online learning and stochastic optimization. *J. Mach. Learn. Res.* **2011**, *12*, 2121–2159.

4. Hardt, M.; Recht, B.; Singer, Y. Train faster, generalize better: Stability of stochastic gradient descent. In Proceedings of the 33rd International Conference on Machine Learning, New York, NY, USA, 19–24 June 2016; pp. 1225–1234.
5. Liu, L.; Jiang, H.; He, P.; Chen, W.; Liu, X.; Gao, J.; Han, J. On the variance of the adaptive learning rate and beyond. *arXiv* **2019**, arXiv:1908.03265.
6. Ge, R.; Huang, F.; Jin, C.; Yuan, Y. Escaping from saddle points—Online stochastic gradient for tensor decomposition. In Proceedings of the 28th Conference on Learning Theory, Paris, France, 2–6 July 2015; pp. 797–842.
7. Alain, G.; Roux, N.L.; Manzagol, P.A. Negative eigenvalues of the hessian in deep neural networks. *arXiv* **2019**, arXiv:1902.02366.
8. Grosse, R.; Martens, J. A kronecker-factored approximate fisher matrix for convolution layers. In Proceedings of the 33rd International Conference on Machine Learning, New York, NY, USA, 20–22 June 2016; pp. 573–582.
9. Martens, J.; Grosse, R. Optimizing neural networks with kronecker-factored approximate curvature. In Proceedings of the 32nd International Conference on Machine Learning, Lille, France, 6–14 July 2015; pp. 2408–2417.
10. Martens, J.; Ba, J.; Johnson, M. Kronecker-factored curvature approximations for recurrent neural networks. In Proceedings of the 6th International Conference on Learning Representations, Vancouver, BC, Canada, 30 April–3 May 2018; pp. 1–25.
11. Martens, J. Deep learning via hessian-free optimization. In Proceedings of the 27th International Conference on International Conference on Machine Learning, Haifa, Israel, 21–24 June 2010; pp. 735–742.
12. Xu, P.; Roosta, F.; Mahoney, M.W. Newton-type methods for non-convex optimization under inexact hessian information. *Math. Program.* **2020**, *184*, 35–70. [CrossRef]
13. Kiros, R. Training neural networks with stochastic hessian-free optimization. *arXiv* **2013**, arXiv:1301.3641.
14. Martens, J.; Sutskever, I. Training deep and recurrent networks with hessian-free optimization. In *Neural Networks: Tricks of the Trade*; Springer: Berlin/Heidelberg, Germany, 2012; pp. 479–535.
15. Botev, A.; Ritter, H.; Barber, D. Practical gauss-newton optimisation for deep learning. In Proceedings of the 2017 International Conference on Machine Learning, Sydney, Australia, 6–11 August 2017; pp. 557–565.
16. George, T.; Laurent, C.; Bouthillier, X.; Ballas, N.; Vincent, P. Fast approximate natural gradient descent in a kronecker-factored eigenbasis. *arXiv* **2018**, arXiv:1806.03884.
17. Amari, S.i.; Park, H.; Fukumizu, K. Adaptive method of realizing natural gradient learning for multilayer perceptrons. *Neural Comput.* **2000**, *12*, 1399–1409. [CrossRef] [PubMed]
18. Pascanu, R.; Bengio, Y. Revisiting natural gradient for deep networks. *arXiv* **2013**, arXiv:1301.3584.
19. Bottou, L.; Curtis, F.E.; Nocedal, J. Optimization methods for large-scale machine learning. *Siam Rev.* **2018**, *60*, 223–311. [CrossRef]
20. Ren, Y.; Goldfarb, D. Efficient subsampled Gauss–Newton and natural gradient methods for training neural networks. *arXiv* **2019**, arXiv:1906.02353.
21. Cai, T.; Gao, R.; Hou, J.; Chen, S.; Wang, D.; He, D.; Zhang, Z.; Wang, L. A gram-gauss-newton method learning overparameterized deep neural networks for regression problems. *arXiv* **2019**, arXiv:1905.11675.
22. Goldfarb, D.; Ren, Y.; Bahamou, A. Practical quasi-Newton methods for training deep neural networks. *arXiv* **2020**, arXiv:2006.08877.
23. Ren, Y.; Goldfarb, D. Kronecker-factored Quasi–Newton Methods for Convolutional Neural Networks. *arXiv* **2021**, arXiv:2102.06737.
24. Byrd, R.H.; Hansen, S.L.; Nocedal, J.; Singer, Y. A stochastic quasi-Newton method for large-scale optimization. *SIAM J. Optim.* **2016**, *26*, 1008–1031. [CrossRef]
25. Wang, X.; Ma, S.; Goldfarb, D.; Liu, W. Stochastic quasi-Newton methods for nonconvex stochastic optimization. *SIAM J. Optim.* **2017**, *27*, 927–956. [CrossRef]
26. Goodfellow, I.; Bengio, Y.; Courville, A.; Bengio, Y. *Deep Learning*; MIT Press: Cambridge, MA, USA, 2016.
27. Horn, R.A.; Johnson, C.R. *Matrix Analysis*; Cambridge University Press: Cambridge, UK, 2012.
28. Steeb, W.H.; Shi, T.K. *Matrix Calculus and Kronecker Product with Applications and C++ Programs*; World Scientific: Singapore, 1997.

Article

A Cascade Deep Forest Model for Breast Cancer Subtype Classification Using Multi-Omics Data

Ala'a El-Nabawy [1,†], Nahla A. Belal [2,3,*,†] and Nashwa El-Bendary [2,3,†]

1. Orange Labs., Smart Village 12511, Giza Governorate, Egypt; aelnabwy@gmail.com
2. College of Computing and Information Technology, Arab Academy for Science, Technology, and Maritime Transport, Smart Village, Giza 12577, Egypt; nashwa.elbendary@aast.edu
3. College of Computing and Information Technology, Arab Academy for Science, Technology, and Maritime Transport, Aswan 81531, Egypt
* Correspondence: nahlabelal@aast.edu
† These authors contributed equally to this work.

Citation: El-Nabawy, A.; Belal, N.A.; El-Bendary, N. A Cascade Deep Forest Model for Breast Cancer Subtype Classification Using Multi-Omics Data. *Mathematics* **2021**, *9*, 1574. https://doi.org/10.3390/math9131574

Academic Editor: Snezhana Gocheva-Ilieva

Received: 18 May 2021
Accepted: 27 June 2021
Published: 4 July 2021

Publisher's Note: MDPI stays neutral with regard to jurisdictional claims in published maps and institutional affiliations.

Copyright: © 2021 by the authors. Licensee MDPI, Basel, Switzerland. This article is an open access article distributed under the terms and conditions of the Creative Commons Attribution (CC BY) license (https://creativecommons.org/licenses/by/4.0/).

Abstract: Automated diagnosis systems aim to reduce the cost of diagnosis while maintaining the same efficiency. Many methods have been used for breast cancer subtype classification. Some use single data source, while others integrate many data sources, the case that results in reduced computational performance as opposed to accuracy. Breast cancer data, especially biological data, is known for its imbalance, with lack of extensive amounts of histopathological images as biological data. Recent studies have shown that cascade Deep Forest ensemble model achieves a competitive classification accuracy compared with other alternatives, such as the general ensemble learning methods and the conventional deep neural networks (DNNs), especially for imbalanced training sets, through learning hyper-representations through using cascade ensemble decision trees. In this work, a cascade Deep Forest is employed to classify breast cancer subtypes, IntClust and Pam50, using multi-omics datasets and different configurations. The results obtained recorded an accuracy of 83.45% for 5 subtypes and 77.55% for 10 subtypes. The significance of this work is that it is shown that using gene expression data alone with the cascade Deep Forest classifier achieves comparable accuracy to other techniques with higher computational performance, where the time recorded is about 5 s for 10 subtypes, and 7 s for 5 subtypes.

Keywords: METABRIC dataset; breast cancer subtyping; deep forest; multi-omics data

1. Introduction

Breast cancer is one of the main causes of cancer death worldwide. Computer-aided diagnosis systems aim to reduce the cost of diagnosis while maintaining the same efficiency of the process. Conventional classification methods depend on feature extraction methods, and to overcome many difficulties of those feature-based methods, deep learning techniques are becoming important approaches to adopt.

Breast cancer classifiers use different methods and different data. Some methods use images [1–3], some use biological data [4,5], and some integrate many types of data [6,7].

Many recent studies have incorporated deep learning, and especially Deep Forest in their studies. Deep forest is still a young research area. However, a lot of work has shown promising results for employing this model in healthcare systems and bioinformatics. For example, reference [8] presents GcForest-PPI, which is a model that uses Deep Forest for the prediction of protein–protein interaction networks. Their model showed and enhanced prediction accuracy and a suggested improvement in drug discovery. The work in reference [9] combined Deep Forest and autoencoders, for the prediction of lncRNA-miRNA interaction, and their model showed improved results. Additionally, reference [10] uses deep learning with Random Forests on the METABRIC dataset, to make use of the different types of data. Their results enhanced the sensitivity values by 5.1%. Additionally, several studies have used deep learning and Deep Forest with the histopathological images data and mammography images [2,3,11–13].

IntClust is breast cancer subtyping technique into 10 subtypes. The IntClust subtyping is dependent on molecular drivers that are obtained using combined genomics and transcriptomic data. Pam50 is another breast subtyping method, and it consists of 5 subtypes [14].

For breast cancer subtyping, several data combined with several techniques have been employed. Since 2011, Mendes et al. [15] employed a clustering method with gene expression data, and showed that the subtyping obtained confirms with already established subtypes. Gene expression and methylation data have been used with different Random Forests models [16]. The study showed that gene expression data outperforms methylation data; however, some features are only discovered using methylation. The work done in [17] uses histopathological images, which was covered above; however, this work is specific to breast cancer. Histopathological images were used with a Stacked Sparse Autoencoder (SSAE), which is a deep learning strategy, and has shown improved performance, with an F-measure of 84.49% and an average area under Precision-Recall curve (AveP) 78.83%. Reference [18] proposes a method that uses histopathological images and extracts features using a convolutional neural network (CNN). The CNN designed obtained an enhanced performance, which was also slightly better when different CNNs were fused. In 2017, Bejnordi et al. [19] applied deep learning algorithms to detect lymph nodes for breast cancer in whole-slide pathology images cans and proposed an improved diagnosis.

Deep forest was used in [20] to classify cancer subtypes, and the model suffered from overfitting and ensemble diversity challenges because of small sample size and high dimensionality of biological data. This is overcome in the use of extensive biological data in this research by employing the METABRIC dataset.

A deep learning technique has also been proposed in [21], and it shows a higher performance than traditional machine learning methods for cancer subtype classification.

The small data size and imbalanced data problems have been addressed in [22], where an enhanced algorithm to handle the data was proposed, combining traditional techniques with deep learning methods. The results obtained confirmed that deep learning enhances performance; in addition to that, methylation data were suggested to be effectively used to improve diagnosis of cancer.

In reference [5], a deep neural networks model uses multi-omics data to classify breast cancer subtypes. The types of omics data used were mRNAdata, DNAmethylation data, and copy number variation (CNV), and the system achieved higher accuracy and area under curve.

Additionally, the authors in reference [6] confirm that deep neural networks perform better than traditional methods as it automatically extracts features from raw data. The data used is copy number alteration and gene expression data for breast cancer patients (METABRIC). The model presented integrates the datasets and the performance is superior to other models.

Moreover, the authors in reference [23] use a network propagation method with a deep embedded clustering (DEC) method to classify the breast tumors into four subtypes. Reference [24] employs deep learning techniques for feature extraction and classification to classify breast cancer lesions using mammograms. The system achieved high accuracy using fused deep features for two datasets compared to similar methods. Additionally, Zhang et al. [25] used a convolutional neural network (CNN) and a recurrent neural network (RNN) to classify three breast cancer subtypes using MRI data. The accuracy achieved was 91% and 83%, using CNN and CLSTM, respectively. Reference [26] combined graph convolutional network (GCN) and convolutional neural network (CNN) to analyze breast mammograms with an accuracy of 96.10 ± 1.60%. Reference [27] also used deep learning on histopathological images for breast cancer subtyping. Deep feature fusion and enhanced routing (FE-BkCapsNet) is used, and results achieved over 90% of accuracy.

Among many others, references [4,7] integrated multiple datasets to address the problem of cancer subtype classification. Xue et al. [4] used integrated omics data for cancer subtype classification using a deep neural forest model, HI-DFNForest, proposing an improved performance. It has been shown that integration of multi-omics data may enhance cancer subtype classification. However, not all types of data are available extensively, and not all types of data add to the classification process. In addition, employing all types of data imposes the restriction of very high time requirements. In the process of diagnosis of cancer, time becomes an important issue due to the critical cases of patients. In reference [7], it was shown that without using sampling techniques on the METABRIC dataset, the results obtained for classification of cancer subtypes are low. Additionally, there is a very big challenge regarding images, where the number of available samples is only 208, much lower than other omics data. Moreover, the techniques used to achieve the highest accuracy obtained were relatively high. In this paper, it was shown that gene expression alone can achieve comparable results on the Deep Forest configuration employed. In addition to achieving very fast performance.

The objective of this work is to extend the previous research [7] by employing a Deep Forest model for using feature combining and classifying the generated integrative data profiles, and enhancing the previously proposed framework through using the full dimension gene expression data and examining the computational performance.

In this paper, a cascade Deep Forest is employed to classify breast cancer subtypes for both subtyping, IntClust (10 subtypes) and Pam50 (5 subtypes), using the METABRIC datasets, namely clinical, gene expression, CNA, and CNV. The full dataset is used, without dimensionality reduction, and without sampling. Several configurations for the cascade Deep Forest are employed and the results obtained are an accuracy of 83.45% for 5 subtypes and 77.55% for 10 subtypes. Other obtained performance metrics also confirm the outperformance of employing gene expression solely, where the precision, recall, specificity, F1-measure, Jaccard, Hamming loss, and Dice are 0.822, 0.774, 0.961, 0.772, 0.640, 0.225, 0.709, respectively, for 10 subtypes. The measures are 0.8421, 0.833, 0.904, 0.820, 0.711, 0.166, and 0.852, for the 5 subtypes, respectively. The precision and Dice measures are slightly higher for the integrated profile gene expression, clinical, CNA, and CNV. The significance of this work is that it is shown that using gene expression data alone with the cascade Deep Forest classifier achieves comparable accuracy to other techniques with higher computational performance, where the time recorded is about 5 s for 10 subtypes, and 7 s for 5 subtypes.

The main contribution of this paper is to:

- Employ the omics METABRIC sub-datasets of gene expression, CNA, and CNV, in addition to the clinical dataset in full dimension without sampling.
- Develop a cascade Deep Forest-based model for breast cancer subtype classification using multi-omics data.
- Obtain comparable results using only omics data without using histopathological images.
- Improve the classification time for breast cancer subtyping through using the cascade Deep Forest classifier.

The rest of this paper is organized as follows. The methods used in evaluating the employed model for breast cancer subtyping are elaborated in Section 2. In Section 3, experimental results are presented, followed by a discussion in Section 4. Finally, a conclusion is presented in Section 5.

2. Materials and Methods

The proposed system in this manuscript uses integrative clinical data and genomics data generated from the extraction and combination of the gene expression, Copy Number Aberrations (CNA), and Copy Number Variations (CNV) feature sets from the genomics dataset.

As depicted in Figure 1, the proposed approach is composed of 4 phases; namely (1) Data acquisition of METABRIC breast cancer subtypes datasets, (2) Data preparation and preprocessing, (3) Integrated data profiles generation, and (4) Cascade Deep Forest-based classification.

After the first phase of four breast cancer subtypes datasets acquisition, the proposed system moves to the second phase of data preparation and preprocessing with only three sub-datasets; namely the clinical data, the features of Copy Number Aberrations (CNA) and Copy Number Variations (CNV) data types, as the fourth sub-dataset of gene expression is submitted as it is without any preprocessing to the third phase of integrated data profiles generation. In the second phase, data cleaning and imputation preprocessing are applied to the clinical data, whereas statistical analysis is applied to the CNA and CNV features. Subsequently, in phase three, the data profiles are generated by concatenating the genomics and clinical features to obtain the integrated data profiles. Finally, the stages of classification process are employed in the fourth phase for training and teasing the proposed system through using the cascade Deep Forest model. The following subsections explain each phase in more details.

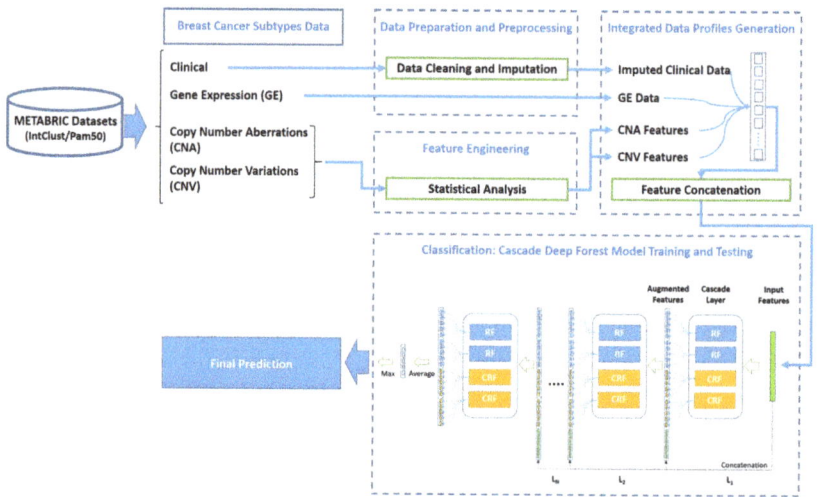

Figure 1. General structure of the proposed approach.

2.1. Data Acquisition

In this phase the breast cancer subtypes dataset used in the conducted experiments is the METABRIC dataset that contains several sub-datasets. The datasets considered for the research conducted in this manuscript are: clinical dataset, gene expression dataset, Copy Number Aberrations (CNA) dataset, and Copy Number Variations (CNV) dataset. The source of the gene expression, CNA and CNV data are the European Genome-phenome Archive (EGA) platform with the accession number, DAC ID, EGAC00001000484 [28]. However, the clinical data are obtained from the Synapse platform [29]. The datasets obtained contain datasets for validation and discovery.

The clinical data available is categorized into four main categories of 27 features. First, personal, which contains only the age at diagnosis. Second, the clinical pathology data, which is data about the tumor, including, size, lymph nodes data, grade, histological type, different hormonal levels, and other features. Third, the treatment category, which indicates the type of treatment received by the patient. Fourth, survival features, which are the status and time.

The Copy Number Aberration (CNA) dataset contains a total of 13 features describing chromosome regions, namely information in somatic tissues about the markers count and mutation type. The dataset also contains information about location, including five features.

In addition to, information about the number of genes in each segment and mutation type described in seven features. Similarly, the Copy Number Variation (CNV) dataset contains 13 features describing chromosome regions in germline tissues about the markers number, mutation type, location, and genes count.

The gene expression dataset contains 48,803 genes expressed using Illumina Sequenced HT 12 array v3.

At the end of breast cancer subtypes data acquisition phase, each obtained dataset is submitted to a data preparation and preprocessing module in phase 2.

2.2. Data Preparation and Preprocessing

This section presents a discussion for the datasets preparation and preprocessing.

2.2.1. Data Preparation

The METABRIC dataset was obtained as explained earlier, and the discovery part was extracted with its labels to include the clinical dataset, gene expression, CNA, and CNV sets. Each of the resulting datasets was prepared according to the following steps:

1. Submitting the CNA and CNV feature files to the statistical analysis feature engineering stage in the data preprocessing stage.
2. Submitting the clinical dataset to the preprocessing stage for data cleaning and imputation.
3. Transposing the gene expression data using Equation (1), then submitting it without any preprocessing to the third phase of the proposed system, where $(A)_{ji}$ represents the matrix of the original gene expression data and $(A^T)_{ij}$ represents the resulted transposed matrix.

$$(A^T)_{ij} = (A)_{ji} \quad \forall i, j. \tag{1}$$

2.2.2. Data Preprocessing

Following the preparation of data, the clinical dataset is submitted to the preprocessing stage for data cleaning and imputation. In this stage, a statistical analysis feature engineering scheme is applied to the CNA and CNV datasets. For that scheme, the frequency and the actual segment percentages Amplification (AMP), Insertions (GAIN), Homozygous Deletion (HOMD), Heterozygous Deletion (HETD) and Neutral (NEUT) were calculated for each chromosome. Figure 2 shows the detailed steps for statistical feature engineering of CNA and CNV data.

The clinical dataset features are encoded using textual categorical encoding. First, features with missing values more than 50% (like NOT_IN_OSLOVAL_P53_mutation_type), and features with 90% blank values (like NOT_IN_OSLOVAL_P53_mutation_details) are deleted. Data imputation [30] is performed on other features missing values with lower ratios. On the other hand, the gene expression dataset is submitted to the integrated data profiles generation phase as it is, without any preprocessing.

2.3. Integrated Data Profiles Generation

At the end of phase 2, the resulted CNA and CNV statistical feature sets, imputed clinical, and the features of gene expression datasets are submitted to phase 3 of integrated data profiles generation, to be concatenated and then generate the output set of integrated data profiles.

2.4. Cascade Deep Forest Based Classification

During phase 4 of the proposed system and after generating integrated data profiles through feature concatenation of imputed clinical data, GE data, and statistically engineered CNA and CNV data, a cascade Deep Forest model is applied for cancer subtype classification. As a preparatory step before the classification phase, the features obtained in phase 3 are split into subsets of 2/3 for training and 1/3 for testing.

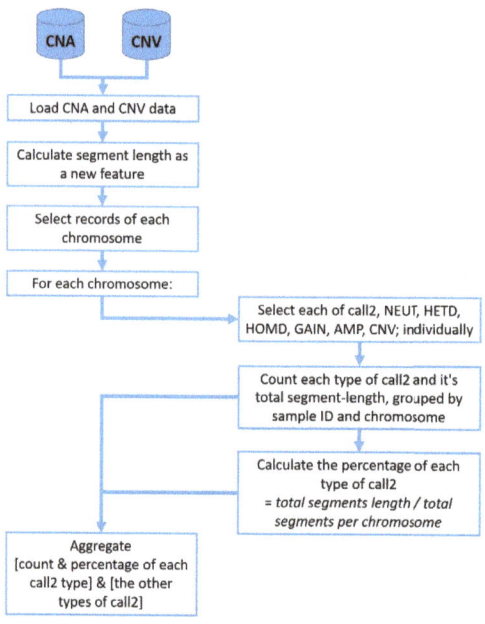

Figure 2. Statistical feature engineering of CNA and CNV data.

The motivation for considering cascade Deep Forest model in the proposed breast cancer subtype classification approach is that conventional supervised machine learning classifiers typically work with labeled data as well as neglecting a considerable amount of data with insufficient information. Consequently, small sample size of training data limits the progress in designing appropriate classifiers. Moreover, several challenges may limit the application of common conventional machine leaning models, such as Support Vector Machine (SVM) and Random Forests, to the task of cancer subtype classification. The sounding challenge is strengthening the risk of overfitting in training, which is characterized by using small sample size and high dimensionality of biology multi-omics data. Additionally, class-imbalance is a very common situation in multi-omics data, which augments the difficulties of model learning with the risk of weakening the ability of model estimation for large sequencing bias. Although several approaches have been recently developed to address the stated challenges [31,32], limited alternatives are proposed with validated methods for small-scale multi-omics data. Additionally, more accurate and robust methods still need further developments for achieving accurate of breast cancer subtype classification.

On the other hand, compared to the typical architecture of convolutional deep neural networks (DNNs) with several convolutional layers and fully connected layers, the DNNs are also highly prone to overfitting, with more chances for convergence to local optimums, when providing imbalanced or relatively small-size training data. However, dropout and regularization methods are widely applied to alleviate that problem, overfitting is still an inescapable problem for DNNs. Thus, the state-of-the-art recommended the cascade Deep Forest model as an efficient alternative to DNNs for learning hyper-level representations in more optimized way.

The cascade Deep Forest model fully uses the characteristics of both deep neural networks and ensemble models. The cascade Deep Forest learn features of class distribution by assembling decision tree-based forests while supervising the input, rather than the overhead of applying forward and backward propagation algorithms to learn hidden variables as in deep neural networks [33].

The cascade forest follows a supervised learning scheme based on layers, which employs ensemble Random Forests to obtain a class distribution of features that results in more precise classification [20,34]. The feature importance in the cascade Deep Forest model is not taken into account among multiple layers during the feature representation training. Accordingly, the prediction accuracy obtained is highly affected by the number of decision trees in each forest, especially with small-scale or imbalanced data, as it is critical in the construction of decision trees, where the discriminative features are used to decide splitting nodes. Figure 3 shows the architecture of the employed cascade forest.

As illustrated in Figure 3, considering the used cascade Deep Forest model, each level of the cascade consists of two Random Forests (RF) (the blue blocks) and two Completely Random Forests (CRF) (the yellow blocks). Therefore, suppose there are n subclasses to predict, then each forest should output an n-dimensional class vector, which is then concatenated for representing the original input.

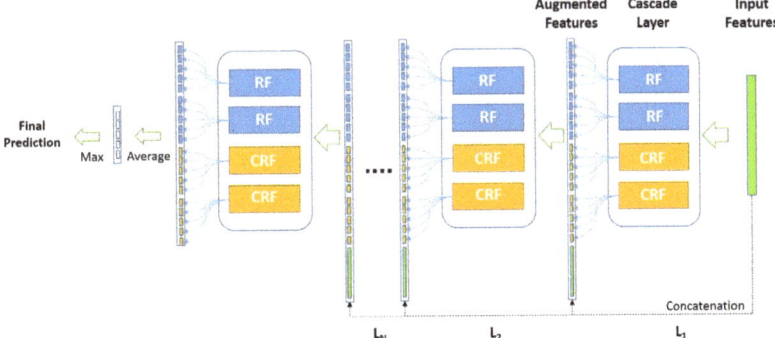

Figure 3. General structure of the cascade Deep Forest network.

3. Results

This section shows the results for different Deep Forest configurations against the 10 subtypes (IntClust) and the 5 subtypes (Pam50). The results in this paper are obtained using the full dimension dataset of the gene expression. The experiments use the whole 48,803 features set. Initially, the Deep Forest configuration was used with the dimensionally reduced gene expression data and the accuracy results were as low as 27%. This led to using the whole gene expression dataset of 48,803 features and the performance obtained showed promising results with high configurations. The performance is shown relative to the time taken per each run. Different Deep Forest configurations were used against the 10 and 5 subtypes. The number of estimators is increased to 900 to make sure that there is no increase in the accuracy. It is not easy to decide the most fitting configuration. For the 10 subtypes, the accuracy reached is 77.55% for the gene expression dataset using 100 trees in each forest, 100 estimators, 5 layers, and 10 k-folds. with time 5:08 s. For the 5 subtypes, the highest accuracy is achieved for the gene expression data and relatively for the CNV and CNA, using 300 trees in each forest, 300 estimators, 5 layers, and 5 k-folds. This accuracy is 83.45% for gene expression, with approximately 55% of accuracy for CNV and CNA, which is comparable to other configurations regarding time. However, performance was not the highest for the clinical data. The time achieved is 7:53 s.

Both experiments were performed using different number of trees per forest and different number of estimators. Specifically, 100, 300, 500, 700 for trees and estimators, using also 900 for the 5 subtypes. Those numbers were used once with 10 layers with 10 k-fold and once with 5 layers with 5 k-fold. To be more confident about the most fitting architecture, another run using 5 layers with 10 K-fold and 10 layers with 5 k-folds was performed. An extra experiment for gene expression data was performed using another combination of layers and k-fold to confirm the results, using 5 layers with 5 k-fold and 10 layers with 10 k-fold.

3.1. Results for Breast Cancer 10 Subtypes

Tables 1–4, show the results for 10 subtypes using all datasets, except images, since the number of samples in the images dataset is much lower than other datasets. The tables show that for gene expression, the highest accuracy achieved is 77.55%, with 100 trees, 100 estimators, 5 layers, and 10 k-folds, with time of 5:08 s. For clinical data, the highest accuracy achieved is 44.22%, with 100 trees, 100 estimators, 5 layers, and 5 k-folds, with time of 0:41 s. For CNV and CNA, the accuracy achieved is 55.78% and 53.40%, respectively, with 500 trees, 500 estimators, 10 layers, 10 k-folds, with a time of 23:01 and 22:40 s, respectively.

Table 1. Gene Expression—10 Subtypes.

Trees/Estimators	Layers/k-Fold	Training Accuracy	Testing Accuracy	Duration
100/100	5/5	70.61%	73.13%	2:50
100/100	10/5	71.04%	74.83%	5:57
100/100	5/10	70.76%	77.55%	5:08
100/100	10/10	71.75%	73.47%	17:29
300/300	5/5	69.04%	73.81%	4:56
300/300	5/10	68.82%	72.11%	11:46
300/300	10/5	71.33%	75.85%	15:04
300/300	10/10	72.47%	75.85%	28:11
500/500	5/5	67.33%	70.07%	10:34
500/500	5/10	67.73%	70.41%	18:01
500/500	10/5	69.90%	72.11%	23:52
500/500	10/10	69.90%	73.47%	67:56
700/700	5/5	67.33%	69.05%	9:37
700/700	5/10	67.33%	70.07%	24:29
700/700	10/5	69.47%	72.45%	33:32
700/700	10/10	69.90%	71.77%	69:58

Table 2. Clinical Data—10 Subtypes.

Trees/Estimators	Layers/k-Fold	Training Accuracy	Testing Accuracy	Duration
100/100	5/5	39.66%	44.22%	0:41
100/100	10/10	41.80%	41.16%	3:14
300/300	5/5	39.94%	42.18%	2:20
300/300	10/10	41.374%	38.78%	11:24
500/500	5/5	40.80%	43.20%	2:43
500/500	10/10	41.94%	38.10%	28:22
700/700	5/5	40.51%	43.54%	3:54
700/700	10/10	41.94%	37.07%	52:01

Table 3. CNV Data—10 Subtypes.

Trees/Estimators	Layers/k-Fold	Training Accuracy	Testing Accuracy	Duration
100/100	5/5	54.92%	53.06%	00:52
100/100	10/10	56.49%	54.08%	3:23
300/300	5/5	53.21%	52.04%	1:50
300/300	10/10	57.49%	55.10%	9:11
500/500	5/5	54.78%	52.38%	2:56
500/500	10/10	56.63%	55.78%	23:01
700/700	5/5	55.06%	52.03%	3:24
700/700	10/10	57.06%	54.42%	38:10

Table 4. CNA Data—10 Subtypes.

Trees/Estimators	Layers/k-Fold	Training Accuracy	Testing Accuracy	Duration
100/100	5/5	52.64%	52.04%	00:44
100/100	10/10	55.92%	51.70%	4:27
300/300	5/5	55.35%	52.38%	1:48
300/300	10/10	56.06%	53.06%	9:37
500/500	5/5	53.92%	52.38%	2:32
500/500	10/10	56.35%	53.40%	22:44
700/700	5/5	53.64%	52.04%	3:25
700/700	10/10	55.92%	53.06%	42:18

Tables 5 and 6, show results of different integrated data profiles for the IntClust (10 subtypes). The highest accuracy achieved for the 10 subtypes is 75.85%, for integrating clinical data with gene expression; however, gene expression alone still achieves a higher accuracy, as reported earlier. Please note that the accuracy reported in Table 6 is the overall accuracy, while the accuracy in the other tables is the reached max layer accuracy. This is the reason for the slight variation in the accuracy values. Moreover, the precision, recall, specificity, F1-measure, Jaccard, Hamming loss, and Dice are 0.822, 0.774, 0.961, 0.772, 0.640, 0.225, 0.709, respectively. The results of the obtained performance and statistical measures confirm the superiority of employing only gene expression.

Table 5. Integrated Profiles Classification Accuracy for 10 Subtypes.

Integrated Profile	Training Accuracy	Testing Accuracy
GE	70.76%	77.55%
Clinical	40.51%	43.54%
CNA	53.21%	51.70%
CNV	55.06%	53.06%
GE + Clinical	67.76%	73.47%
Clinical + GE	71.61%	75.85%
GE + CNV	69.76%	76.19%
GE + CNA	66.76%	70.75%
Clinical + CNA	58.63%	60.20%
Clinical + CNV	61.06%	57.82%
CNA + CNV	55.21%	51.70%
Clinical + CNV + GE	68.33%	72.79%
Clinical + CNA + GE	68.90%	71.09%
Clinical + CNA + CNV	60.34%	61.56%
GE + CNA + CNV	71.61%	74.83%
GE + CNV + CNA + Clinical	68.62%	74.15%
GE + CNA + CNV + Clinical	70.33%	74.49%
GE + Clinical + CNA + CNV	68.76%	72.79%

Table 6. Integrated Profiles Classification Performance Metrics for 10 Subtypes.

Data Profile	Accuracy	Precision	Recall (Sensitivity)	Specificity	F1-Measure	Jaccard	Hamming-Loss	Dice
GE	77.47%	0.822	0.774	0.961	0.772	0.640	0.225	0.709
Clinical	41.97%	0.369	0.4197	0.919	0.367	0.266	0.580	0.320
CNA	49.48%	0.445	0.494	0.926	0.450	0.309	0.505	0.297
CNV	53.24%	0.554	0.532	0.932	0.491	0.341	0.467	0.340
GE + Clinical	72.7%	0.789	0.726	0.956	0.708	0.573	0.273	0.561
GE + CNV	76.11%	0.812	0.761	0.961	0.749	0.613	0.239	0.644
GE + CNA	70.64%	0.749	0.706	0.951	0.683	0.546	0.293	0.555
Clinical + CNA	58.02%	0.5286	0.5802	0.938	0.541	0.402	0.419	0.522
Clinical + CNV	57.6%	0.521	0.576	0.938	0.537	0.3959	0.423	0.44
CNA + CNV	51.54%	0.472	0.515	0.9303	0.463	0.323	0.484	0.217
Clinical + CNV + GE	69.28%	0.753	0.693	0.950	0.676	0.532	0.307	0.611
Clinical + CNA + GE	70.9%	0.767	0.7098	0.953	0.6919	0.552	0.2901	0.623
CNA + CNV + Clinical	59.01%	0.557	0.6143	0.939	0.578	0.4387	0.3856	0.590
GE + CNA + CNV	74.74%	0.809	0.747	0.958	0.732	0.599	0.252	0.678
GE + CNV + CNA + Clinical	74.06%	0.809	0.7406	0.955	0.723	0.590	0.259	0.621

3.2. Results for Breast Cancer 5 Subtypes

For the 5 subtypes, Tables 7–10 show results of different datasets. For gene expression, the highest accuracy achieved is 83.45%, with 300 trees, 300 estimators, 5 layers, and 5 k-folds, with time of 7:53 s. For clinical data, the highest accuracy achieved is 76.35%, with 700 trees, 700 estimators, 10 layers, and 10 k-folds, with time of 16:28 s. For CNV, the accuracy achieved is 56.76%, with 500 trees, 500 estimators, 10 layers, 10 k-folds, with a time of 12:04 s. For CNA, the accuracy achieved is 56.42%, with 700 trees, 700 estimators, 5 layers, 5 k-folds, with a time of 3:54 s.

Table 7. Gene Expression—5 Subtypes.

Trees/Estimators	Layers/k-Fold	Training Accuracy	Testing Accuracy	Duration
100/100	5/5	81.26%	80.07%	1:59
100/100	5/10	81.12%	79.39%	8:50
100/100	10/5	81.25%	82.09%	7:11
100/100	10/10	81.69%	81.76%	13:25
300/300	5/5	83.55%	83.45%	7:53
300/300	5/10	80.26%	80.07%	13:09
300/300	10/5	81.83%	79.73%	18:05
300/300	10/10	83.12%	81.42%	37:08
500/500	5/5	78.40%	80.07%	9:03
500/500	5/10	79.69%	80.41%	23:13
500/500	10/5	81.69%	81.08%	22:56
500/500	10/10	81.83%	82.77%	42:41
700/700	5/5	78.97%	79.73%	9:01
700/700	5/10	79.11%	80.07%	23:03
700/700	10/5	81.40%	81.08%	29:40
700/700	10/10	81.83%	82.77%	68:15
900/900	5/5	78.54%	79.73%	11:40
900/900	5/10	78.97%	80.41%	29:32
900/900	10/5	81.69%	82.77%	38:02
900/900	10/10	81.97%	82.43%	72:10

Table 8. Clinical Data—5 Subtypes.

Trees/Estimators	Layers/k-Fold	Training Accuracy	Testing Accuracy	Duration
100/100	5/5	73.46%	74.66%	0:38
100/100	10/10	73.25%	75.34%	4:51
300/300	5/5	72.96%	75.00%	2:25
300/300	10/10	74.11%	74.66%	10:28
500/500	5/5	73.39%	75.00%	2:54
500/500	10/10	73.96%	75.00%	10:33
700/700	5/5	74.25%	75.00%	3:35
700/700	10/10	73.96%	76.35%	16:28
900/900	5/5	74.11%	75.00%	4:09
900/900	10/10	73.82%	74.66%	17:01

Table 9. CNV Data—5 Subtypes.

Trees/Estimators	Layers/k-Fold	Training Accuracy	Testing Accuracy	Duration
100/100	5/5	63.09%	56.08%	0:43
100/100	10/10	63.38%	56.42%	2:46
300/300	5/5	62.37%	55.41%	2:29
300/300	10/10	63.95%	55.41%	9:13
500/500	5/5	63.52%	55.47%	3:10
500/500	10/10	63.38%	56.76%	12:04
700/700	5/5	63.23%	55.41%	3:44
700/700	10/10	64.23%	55.74%	18:00
900/900	5/5	62.95%	55.74%	4:12
900/900	10/10	64.52%	56.76%	19:32

Table 10. CNA Data—5 Subtypes.

Trees/Estimators	Layers/k-Fold	Training Accuracy	Testing Accuracy	Duration
100/100	5/5	63.52%	56.08%	0:52
100/100	10/10	65.09%	54.73%	3:09
300/300	5/5	62.80%	55.74%	3:07
300/300	10/10	64.38%	55.07%	11:27
500/500	5/5	62.95%	55.41%	4:09
500/500	10/10	64.52%	55.41%	13:37
700/700	5/5	62.95%	56.42%	3:54
700/700	10/10	64.38%	55.74%	14:56
900/900	5/5	62.95%	55.74%	4:18
900/900	10/10	64.09%	55.41%	19:19

Tables 11 and 12 show results of different integrated data profiles for the Pam50 (5 subtypes) subtyping. The highest accuracy achieved for the 5 subtypes is 80.41%, for CNA, CNV, gene expression, and clinical data. Still, gene expression alone achieves a higher prediction accuracy. The slight variation in accuracy reported in Tables 6 and 12 is that it is the overall accuracy, while the accuracy in the other tables is the reached max layer accuracy. In addition to the obtained accuracy, precision, recall, specificity, F1-measure, Jaccard, Hamming loss, and Dice are 0.8421, 0.833, 0.904, 0.820, 0.711, 0.166, and 0.852, respectively. The precision and Dice measures are slightly higher for the integrated profile gene expression, clinical, CNA, and CNV. However, the remaining measures are in favor of the gene expression profile, which confirm the outperformance of the suggested data profile.

Table 11. Integrated Profiles Classification Accuracy for 5 Subtypes.

Integrated Profile	Training Accuracy	Testing Accuracy
GE	83.55%	83.45%
Clinical	72.96%	75.00%
CNA	62.80%	55.74%
CNV	62.37%	55.41%
GE + Clinical	81.26%	82.09%
Clinical + GE	80.40%	79.05%
GE + CNV	81.26%	79.05%
GE + CNA	81.12%	79.39%
Clinical + CNA	73.82%	69.93%
Clinical + CNV	74.54%	70.95%
CNA + CNV	63.09%	56.76%
Clinical + CNV + GE	80.11%	78.38%
Clinical + CNA + GE	80.83%	78.72%
Clinical + CNA + CNV	73.68%	68.92%
GE + CNA + CNV	80.83%	79.05%
GE + CNV + CNA + Clinical	81.55%	82.09%
GE + CNA + CNV + Clinical	81.12%	80.41%
GE + Clinical + CNA + CNV	80.54%	80.07%

Table 12. Integrated Profiles Classification Performance Metrics for 5 Subtypes.

Data Profile	Accuracy	Precision	Recall (Sensitivity)	Specificity	F1-Measure	Jaccard	Hamming-Loss	Dice
GE	83.38%	0.8421	0.833	0.904	0.820	0.711	0.166	0.852
Clinical	75.59%	0.761	0.755	0.892	0.748	0.611	0.244	0.796
CNA	55.59%	0.5474	0.555	0.705	0.500	0.360	0.444	0.432
CNV	55.93%	0.553	0.559	0.702	0.503	0.362	0.440	0.382
GE + Clinical	82.03%	0.848	0.8203	0.868	0.797	0.687	0.179	0.857
GE + CNV	79.05%	0.765	0.786	0.846	0.754	0.634	0.213	0.845
GE + CNA	79.39%	0.827	0.793	0.849	0.767	0.644	0.206	0.842
Clinical + CNA	67.7%	0.7009	0.677	0.807	0.654	0.504	0.322	0.644
Clinical + CNV	71.1%	0.725	0.711	0.818	0.694	0.546	0.288	0.783
CNA + CNV	56.76%	0.479	0.566	0.697	0.505	0.366	0.433	0.4
Clinical + CNV + GE	78.6%	0.769	0.786	0.841	0.756	0.635	0.213	0.842
Clinical + CNA + GE	78.9%	0.772	0.789	0.8447	0.759	0.639	0.210	0.842
Clinical + CNA + CNV	68.92%	0.644	0.688	0.809	0.659	0.514	0.311	0.727
GE + CNA + CNV	78.9%	0.768	0.789	0.846	0.759	0.64	0.210	0.842
GE + CNV + CNA + Clinical	82.37%	0.848	0.823	0.872	0.801	0.693	0.176	0.857

4. Discussion

In this work, the experiments first make use of different Deep Forest configurations on each dataset solely. Gene expression alone significantly gave the best performance, where the accuracy was 83.45% for 5 subtypes using 300 estimators, 5 layers, 5 k-folds, and the accuracy was 77.55% for 10 subtypes using 100 estimators, 5 layers, and 10 k-folds. This was concluded after experimenting 100,300,500,700 and 900 estimators across 5, 10 layers and 5, 10 k-folds. The integration of datasets was performed by concatenating the datasets and applying the best configuration of Deep Forest to classify it. The results reached did not give any improvement over the highest accuracy reached using gene expression. However, for the 5 subtypes, the integrated profile CNA + CNV achieved 56.7%, while CNA alone achieved 55.74%, and CNV alone achieved 55.41%. For the 10 subtypes, the clinical data alone achieved 43.5%, CNV alone achieved 53.06%, and the CNA alone achieved 51.70%. The integrated clinical data with the CNA achieved 60.20%, the integrated clinical data with CNV achieved 57.82%, and the integrated clinical data with both CNA and CNV achieved 61.56%.

In the research [7], an accuracy of 88.36% was achieved for IntClust (10 subtypes) subtyping using Linear-SVM. The accuracy achieved was using the data profile of clinical, gene expression, CNA, and CNV datasets. For the Pam50 subtyping (5 subtypes), the accuracy was 97.1% using Linear-SVM and E-SVM classifiers, with all data including histopathological images features. However, the images data are not comprehensive, as they are only available for 208 samples, unlike other data, which are extensively available for all patients in the dataset. Moreover, the time taken to obtain the above-mentioned accuracy is extensive. hence, the Deep Forest used in this paper, makes use of the gene expression data alone to achieve comparable results without using any sampling techniques. In [7], the highest accuracy achieved used SMOTE sampling. The highest accuracy achieved among all different data profiles was 71.35% for IntClust, which was outperformed by the Deep Forest configuration in this paper, achieving 77.55%, and a running time of 5:08 s, which is extensively less than the model proposed in [7]. For the gene expression data alone in [7], the accuracy only reached 46.08%. Similarly, for Pam50 (5 subtypes), gene expression alone achieved 78.85%, while the highest recorded accuracy was 80.66%, without images, for the gene expression and clinical data profiles. However, the proposed Deep Forest configuration achieved up to 83.45% of accuracy and 7:53 s run time.

The current study could be further expanded by examining the technique on more datasets for breast cancer subtyping. Additionally, other deep learning methods could be employed to verify the robustness of using gene expression data.

5. Conclusions

This research proposes a Deep Forest classifier for the IntClust and Pam50 breast cancer subtypes. The experiments are carried out using different combinations of trees and estimators, specifically 100, 300, 500, 700, and 900, as well as layers and k-folds of 5 and 10. Gene expression alone significantly gave the best performance, with an accuracy of

83.45% for 5 subtypes and 77.55% for 10 subtypes, and time about 5 s for 10 subtypes, and 7 s for 5 subtypes. The integration of datasets did not give any improvement, where for the 5 subtypes, CNA and CNV data achieved 56.7%, while CNA alone achieved 55.74%, and CNV alone achieved 55.41%. For the 10 subtypes, the clinical data achieved 43.5%, CNV alone achieved 53.06%, and the CNA alone achieved 51.70%. The integrated clinical data with the CNA achieved 60.20%, the integrated clinical data with CNV achieved 57.82%, and the integrated clinical data with both CNA and CNV achieved 61.56%. It is concluded that using gene expression alone achieves comparable results.

Author Contributions: Conceptualization, A.E.-N., N.A.B. and N.E.-B.; methodology, A.E.-N., N.A.B. and N.E.-B.; software, A.E.-N.; validation, A.E.-N., N.A.B. and N.E.-B.; formal analysis, A.E.-N., N.A.B. and N.E.-B.; investigation, A.E.-N., N.A.B. and N.E.-B.; resources, A.E.-N.; data curation, A.E.-N., N.A.B. and N.E.-B.; writing—original draft preparation, N.A.B. and N.E.-B.; writing—review and editing, N.A.B. and N.E.-B.; visualization, A.E.-N., N.A.B. and N.E.-B.; supervision, N.A.B. and N.E.-B.; project administration, N.A.B. and N.E.-B. All authors have read and agreed to the published version of the manuscript.

Funding: This research received no external funding.

Institutional Review Board Statement: Not applicable.

Informed Consent Statement: Not applicable.

Data Availability Statement: The METABRIC dataset was obtained based on a formal access request from our institution to perform the study on the data. Due to the sensitive information in the dataset, Synapse offers access through a controlled use mechanism and the dataset could be requested through the following link: https://www.synapse.org/$#$!Synapse:syn1688369/wiki/27311 (accessed on 15 May 2021).

Conflicts of Interest: The authors declare no conflict of interest.

References

1. Araujo, T.; Aresta, G.; Castro, E.; Rouco, J.; Aguiar, P.; Eloy, C.; Polonia, A.; Campilho, A. Classification of breast cancer histology images using convolutional neural networks. *PLoS ONE* **2017**, *12*, e0177544. [CrossRef] [PubMed]
2. Pan, X.; Lu, Y.; Lan, R.; Liu, Z.; Qin, Z.; Wang, H.; Liu, Z. Mitosis detection techniques in H&E stained breast cancer pathological images: A comprehensive review. *Comput. Electr. Eng.* **2021**, *91*, 107038.
3. Chouhan, N.; Khan, A.; Shah, J.; Hussnain, M.; Khan, M. Deep convolutional neural network and emotional learning based breast cancer detection using digital mammography. *Comput. Biol. Med.* **2021**, *132*, 104318. [CrossRef] [PubMed]
4. Xu, J.; Wu, P.; Chen, Y.; Meng, Q.; Dawood, H.; Dawood, H. A hierarchical integration deep flexible neural forest framework for cancer subtype classification by integrating multi-omics data. *BMC Bioinform.* **2019**, *20*, 527. [CrossRef] [PubMed]
5. Lin, Y.; Zhang, W.; Cao, H.; Li, G.; Du, W. Classifying Breast Cancer Subtypes Using Deep Neural Networks Based on Multi-Omics Data. *MDPI Genes* **2020**, *11*, 888. [CrossRef] [PubMed]
6. Mohaiminul Islam, M.; Huang, S.; Ajwad, R.; Chi, C.; Wang, Y.; Hu, P. An integrative deep learning framework for classifying molecular subtypes of breast cancer. *Comput. Struct. Biotechnol. J.* **2020**, *18*, 2185–2199. [CrossRef] [PubMed]
7. El-Nabawy, A.; El-Bendary, N.; Belal, N.A. A feature-fusion framework of clinical, genomics, and histopathological data for METABRIC breast cancer subtype classification. *Appl. Soft Comput.* **2020**, *91*, 106238. [CrossRef]
8. Yu, B.; Chen, C.; Wang, X.; Yu, Z.; Ma, A.; Liu, B. Prediction of protein–protein interactions based on elastic net and deep forest. *Expert Syst. Appl.* **2021**, *176*, 114876. [CrossRef]
9. Wang, W.; Guan, X.; Khan, M.; Xiong, Y.; Wei, D.Q. LMI-DForest: A deep forest model towards the prediction of lncRNA-miRNA interactions. *Comput. Biol. Chem.* **2020**, *89*, 107406. [CrossRef]
10. Arya, N.; Saha, S. Multi-modal advanced deep learning architectures for breast cancer survival prediction[Formula presented]. *Knowl. Based Syst.* **2021**, *221*, 106965. [CrossRef]
11. Sirinukunwattana, K.; Raza, S.; Tsang, Y.W.; Snead, D.; Cree, I.; Rajpoot, N. Locality Sensitive Deep Learning for Detection and Classification of Nuclei in Routine Colon Cancer Histology Images. *IEEE Trans. Med Imaging* **2016**, *35*, 1196–1206. [CrossRef] [PubMed]
12. Komura, D.; Ishikawa, S. Machine Learning Methods for Histopathological Image Analysis. *Comput. Struct. Biotechnol. J.* **2018**, *16*, 34–42. [CrossRef]
13. Sohail, A.; Khan, A.; Wahab, N.; Zameer, A.; Khan, S. A multi-phase deep CNN based mitosis detection framework for breast cancer histopathological images. *Sci. Rep.* **2021**, *11*, 1–18. [CrossRef]
14. Ali, H.R.; Rueda, O.M.; Chin, S.F.; Curtis, C.; Dunning, M.J.; Aparicio, S.A.; Caldas, C. Genome-driven integrated classification of breast cancer validated in over 7500 samples. *Genome Biol.* **2014**, *431*, 1–14.

15. Mendes, A. Identification of Breast Cancer Subtypes Using Multiple Gene Expression Microarray Datasets. In Proceedings of the Australasian Joint Conference on Artificial Intelligence, Barcelona, Catalonia, Spain, 16–22 July 2011; pp. 92–101.
16. List, M.; Hauschild, A.C.; Tan, Q.; Kruse, T.A.; Baumbach, J.; Batra, R. Classification of Breast Cancer Subtypes by combining Gene Expression and DNA Methylation Data. *J. Integr. Bioinform.* **2014**, *11*, 1–14. [CrossRef]
17. Xu, J.; Xiang, L.; Liu, Q.; Gilmore, H.; Wu, J.; Tang, J.; Madabhushi, A. Stacked sparse autoencoder (SSAE) for nuclei detection on breast cancer histopathology images. *IEEE Trans. Med Imaging* **2016**, *35*, 119–130. [CrossRef]
18. Spanhol, F.; Oliveira, L.; Petitjean, C.; Heutte, L. Breast cancer histopathological image classification using Convolutional Neural Networks. In Proceedings of the 2016 International Joint Conference on Neural Networks (IJCNN), Vancouver, BC, Canada, 24–29 July 2016; Volume 2016, pp. 2560–2567.
19. Bejnordi, B.; Veta, M.; Van Diest, P.; Van Ginneken, B.; Karssemeijer, N.; Litjens, G.; Van Der Laak, J.; Hermsen, M.; Manson, Q.; Balkenhol, M.; et al. Diagnostic assessment of deep learning algorithms for detection of lymph node metastases in women with breast cancer. *JAMA J. Am. Med. Assoc.* **2017**, *318*, 2199–2210. [CrossRef] [PubMed]
20. Guo, Y.; Liu, S.; Li, Z.; Shang, X. BCDForest: A boosting cascade deep forest model towards the classification of cancer subtypes based on gene expression data. *BMC Bioinform.* **2018**, *19*, 1–13. [CrossRef]
21. Gao, F.; Wang, W.; Tan, M.; Zhu, L.; Zhang, Y.; Fessler, E.; Vermeulen, L.; Wang, X. DeepCC: A novel deep learning-based framework for cancer molecular subtype classification. *Oncogenesis* **2019**, *8*, 527. [CrossRef]
22. Dong, Y.; Yang, W.; Wan, J.; Zhao, J.; Qiang, Y. MLW-gcForest: A Multi-Weighted gcForest Model for Cancer Subtype Classification by Methylation Data. *MDPI Appl. Sci.* **2019**, *9*, 3589. [CrossRef]
23. Rohani, N.; Eslahchi, C. Classifying Breast Cancer Molecular Subtypes by Using Deep Clustering Approach. *Front. Genet.* **2020**, *11*, 1108. [CrossRef] [PubMed]
24. Ragab, D.; Attallah, O.; Sharkas, M.; Ren, J.; Marshall, S. A framework for breast cancer classification using Multi-DCNNs. *Comput. Biol. Med.* **2021**, *131*, 104245. [CrossRef] [PubMed]
25. Zhang, Y.; Chen, J.H.; Lin, Y.; Chan, S.; Zhou, J.; Chow, D.; Chang, P.; Kwong, T.; Yeh, D.C.; Wang, X.; et al. Prediction of breast cancer molecular subtypes on DCE-MRI using convolutional neural network with transfer learning between two centers. *Eur. Radiol.* **2021**, *31*, 2559–2567. [CrossRef] [PubMed]
26. Zhang, Y.D.; Satapathy, S.; Guttery, D.; Górriz, J.; Wang, S.H. Improved Breast Cancer Classification Through Combining Graph Convolutional Network and Convolutional Neural Network. *Inf. Process. Manag.* **2021**, *58*, 102439. [CrossRef]
27. Wang, P.; Wang, J.; Li, Y.; Li, P.; Li, L.; Jiang, M. Automatic classification of breast cancer histopathological images based on deep feature fusion and enhanced routing. *Biomed. Signal Process. Control* **2021**, *65*, 102341. [CrossRef]
28. METABRIC Genomics Dataset, The European Genome-Phenome Archive (EGA). Available online: https://ega-archive.org/dacs/EGAC00001000484 (accessed on 15 April 2020).
29. METABRIC Clinical Dataset, Molecular Taxonomy of Breast Cancer International Consortium. Available online: https://www.synapse.org/#!Synapse:syn1688369/wiki/27311 (accessed on 15 April 2020).
30. Dziura, J.; Post, L.; Zhao, Q.; Fu, Z.; Peduzzi, P. Strategies for dealing with missing data in clinical trials: from design to analysis. *Yale J. Biol. Med.* **2013**, *86*, 343–358. [PubMed]
31. Jahid, M.; Huang, T.; Ruan, J. A personalized committee classification approach to improving prediction of breast cancer metastasis. *Bioinformatics* **2014**, *30*, 1858–1866. [CrossRef]
32. Saddiki, H.; McAuliffe, J.; Flaherty, P. GLAD: A mixed-membership model for heterogeneous tumor subtype classification. *Bioinformatics* **2015**, *30*, 225–232. [CrossRef]
33. Fan, Y.; Qi, L.; Tie, Y. The Cascade Improved Model Based Deep Forest for Small-scale Datasets Classification. In Proceedings of the 2019 8th International Symposium on Next Generation Electronics (ISNE), Zhengzhou, China, 9–10 October 2019; pp. 1–3.
34. Wang, H.; Tang, Y.; Jia, Z.; Ye, F. Dense Adaptive Cascade Forest: A Self Adaptive Deep Ensemble for Classification Problems. 2019. Available online: http://xxx.lanl.gov/abs/1804.10885 (accessed on 15 April 2020).

Article

Self-Expressive Kernel Subspace Clustering Algorithm for Categorical Data with Embedded Feature Selection

Hui Chen [1,2], Kunpeng Xu [3], Lifei Chen [4] and Qingshan Jiang [1,*]

1 Shenzhen Institute of Advanced Technology, Chinese Academy of Sciences, Shenzhen 518055, China; hui.chen1@siat.ac.cn
2 Shenzhen College of Advanced Technology, University of Chinese Academy of Sciences, Shenzhen 518055, China
3 Department of Computer Science, University of Sherbrooke, Sherbrooke, QC J1K 2R1, Canada; Kunpeng.Xu@USherbrooke.ca
4 College of Computer and Cyber Security, Fujian Normal University, Fuzhou 350007, China; clfei@fjnu.edu.cn
* Correspondence: qs.jiang@siat.ac.cn; Tel.: +86-755-8639-2340

Citation: Chen, H.; Xu, K.; Chen, L.; Jiang, Q. Self-Expressive Kernel Subspace Clustering Algorithm for Categorical Data with Embedded Feature Selection. *Mathematics* 2021, 9, 1680. https://doi.org/10.3390/math9141680

Academic Editor: Snezhana Gocheva-Ilieva

Received: 18 June 2021
Accepted: 14 July 2021
Published: 16 July 2021

Publisher's Note: MDPI stays neutral with regard to jurisdictional claims in published maps and institutional affiliations.

Copyright: © 2021 by the authors. Licensee MDPI, Basel, Switzerland. This article is an open access article distributed under the terms and conditions of the Creative Commons Attribution (CC BY) license (https://creativecommons.org/licenses/by/4.0/).

Abstract: Kernel clustering of categorical data is a useful tool to process the separable datasets and has been employed in many disciplines. Despite recent efforts, existing methods for kernel clustering remain a significant challenge due to the assumption of feature independence and equal weights. In this study, we propose a self-expressive kernel subspace clustering algorithm for categorical data (SKSCC) using the self-expressive kernel density estimation (SKDE) scheme, as well as a new feature-weighted non-linear similarity measurement. In the SKSCC algorithm, we propose an effective non-linear optimization method to solve the clustering algorithm's objective function, which not only considers the relationship between attributes in a non-linear space but also assigns a weight to each attribute in the algorithm to measure the degree of correlation. A series of experiments on some widely used synthetic and real-world datasets demonstrated the better effectiveness and efficiency of the proposed algorithm compared with other state-of-the-art methods, in terms of non-linear relationship exploration among attributes.

Keywords: machine learning; categorical data; similarity; feature selection; kernel density estimation; non-linear optimization; kernel clustering

1. Introduction

One of the goals of clustering is to mine the internal structure and characteristics of unlabeled data, which is known as unsupervised learning [1,2]. Real-world applications, i.e., pattern recognition [3], text mining [4], image retrieval [5], and bioinformatics [6], generate unlabeled data. All of these data are not just numerical data but are increasingly categorical data, which are flooding into practical applications. Clustering analysis for categorical data has attracted a great deal of interest from the scientific community. One example is that political philosophy is often measured as liberal, moderate, or conservative. Another example is that breast cancer diagnoses based on a mammograms use the categories normal, benign, probably benign, suspicious, and malignant.

In the past few decades, various clustering algorithms have been proposed [7–11] for numerical data. However, the attributes of categorical data are discrete, and their attribute values come from a limited symbol set. Unlike continuous data, categorical data are unable to produce a mathematical calculation, such as the mean and standard deviation. As a result, algorithms suitable for continuous data cannot be directly used for categorical data. To deal with this disadvantage, researchers have developed some clustering algorithms for categorical data, such as ROCK [12], ScaLable Information Bottleneck (LIMBO) [13,14], MGR [15], DHCC [16], and k-modes type algorithms [17–23]. However, each of these algorithms has its own merits and disadvantages. Even state-of-the-art algorithms have their shortcomings, and they are not effective for all datasets. For instance, ROCK is a non-k-mode agglomerative hierarchical clustering method that uses the conventional Jaccard coefficient to compute the similarity of two samples. However, the Jaccard cannot measure the specific value of the difference; it can only obtain whether the result is the same or not. In addition, the time complexity of this algorithm is high, which is quadratic with the number of objects. LIMBO uses an agglomerative information bottleneck to measure the entities' distance, but is not comprehensive enough to extract data clustering features. The MGR algorithm proposes a mean gain ratio to select cluster attributes. LIMBO and MGR are based on information theory, meaning that they can quickly take into account one related variable, but one only, while ignoring other important feature information. DHCC can analyze multiple correspondence, avoiding a one-to-one similarity calculation. However, this method is sensitive to strange objects and, compared with agglomerative approaches, DHCC is a divisive algorithm with less application. The conventional k-modes algorithm and its variants have been extensively used for categorical data clustering. The distance of the samples was measured by simple matching coefficient (SMC). However, these methods only consider the attributes' mode, while ignoring the statistical information of the data itself. Meanwhile, they can be trapped into local optima and are sensitive to initial clusters and modes. Our numerical experiments even showed that the k-modes algorithm could not identify the optimal clustering results for some particular datasets, regardless of the selection of the initial centers.

To solve the k-modes type algorithms' problems, Chen [24] proposed a probabilistic framework in which the kernel bandwidth was introduced with a soft feature selection scheme so that the cluster center equals to the smoothed frequency estimator for the categories. Feature selection is of great significance to data processing in the era of big data [25,26]. It often involves the process of selecting the most important features representing an object's attributes and then building a learning model in tasks clustering. Feature selection can not only relieve the curse of dimensionality caused by too many attributes but can also retain relevant features, remove irrelevant features, reduce the difficulty of learning tasks, and look for the essential features. Based on evaluation criteria, embedded feature selection methods such as CART [27] not only overcome the low efficiency of the wrapper feature selection method [28–30] but also avoid the disconnection of the filter feature selection method. Algorithms that take a filter-method approach to feature selection, such as Chi-Square [31], information gain [32], gain ratio [33], support vector machine [34,35], ReliefF [36,37], and hybrid ReliefF [38,39], are used in many practical applications. The embedded feature selection approach uses a learning model, so that the feature selection process is automatically integrated with the learner training process. Although several clustering analysis methods employ feature selection [24,40], many of the current approaches have one or more of the following disadvantages: considering all features independently, considering all attributes' importance equally, and lack of an optimization solution.

The kernel clustering method that increases the sample features' optimization process uses the Mercer kernel to map the samples in the input space to the high-dimensional feature space and clusters in the feature space. The kernel clustering method is widely used and is considered superior to classical clustering algorithms in performance. It can distinguish, extract, and enlarge useful features through non-linear mapping, so as to achieve more accurate clustering. Kernel k-means algorithm [41] makes the sample linearly separable (or nearly linearly separable) in kernel space by the "kernel" method. Still, the kernel function is defined for continuous data. Thus it cannot be directly transposed to categorical data and the algorithm based on the assumption that the original features are equally important. Some recent self-expressiveness-based methods [42–44] use subspace self-expressiveness property related to regularization terms. They are also not suitable for categorical data, and they all involve a linear combination of attributes.

In this paper, we view the task of clustering categorical data from a kernel clustering approach and propose a non-linear clustering algorithm for categorical data. The algorithm, named self-expressive kernel subspace clustering for categorical data (SKSCC), is based on the kernel density estimation (KDE) and probability-based similarity measurement. SKSCC not only considers the relationship between attributes in non-linear space but also gives each attribute a feature weight to measure the correlation degree. KDE has been employed in the estimation of probability distribution for categorical data [24,45,46]. This work introduces the self-expressive kernel density estimation (SKDE) in which every attribute has its own bandwidth. It then proposes a new non-linear similarity measurement method for categorical data in which a weight is added for each attribute to determine the importance of the attribute. Therefore, the objective function of the derived clustering algorithm is non-linear. As is commonly accepted, non-linear equations and equalities are not easy to solve. Therefore, we propose an efficient non-linear optimization method to solve the objective function of the clustering algorithm.

In summary, the main contributions of our work are as follows:

- We define the self-expressive kernel density estimation approach, in which the symbols can be expressed by probability that is proportional to the kernel bandwidth, and the cluster center is smoothed to the frequency estimator for the categories;
- We propose a non-linear feature-weighted similarity measurement method that gives consideration to the relationship between the attributes;
- We put forward a non-linear optimization method in kernel subspace. Furthermore, we present the SKSCC, an efficient self-expressive kernel subspace clustering algorithm for categorical data that uses feature selection to choose the important attributes;
- A series of experiments on several synthesis and real-world datasets were conducted to compare the performance of the proposed algorithm. The experimental results show that the proposed algorithm outperforms other algorithms in terms of non-linear relationship exploration among attributes and improves the performance and efficiency of clustering.

The remainder of this paper is organized as follows: Section 2 describes related work. Section 3 introduces the KDE-based similarity for categorical data. In Section 4, the new clustering algorithm is elaborated. Experimental results are analyzed in Section 5. Section 6 presents our conclusions.

2. Related Work

The similarity measure of categorical data is the basis of categorical data analysis. A good clustering algorithm maximizes the similarity within clusters and minimizes the similarity between clusters. Although many researchers have proposed different methods to measure the similarity or dissimilarity of categorical data, none of them have been widely recognized. For numerical data, there are Euclidean distance, vector dot product, and other similar or different degrees of measurement objects. For categorical data, the mean and variance are not defined, and the vector dot product operation is meaningless.

In 1998, Huang [17] proposed the conventional k-modes algorithm, which is a non-weighted feature clustering approach. The k-modes algorithm can be formulated into a mathematical optimization model as follows:

$$\min J(W, Q) = \sum_{l=1}^{k} \sum_{i=1}^{n} w_{li} d(X_i, Q_l) \qquad (1)$$

where w_{li} composes a partition matrix and $\sum_{l=1}^{k} w_{li} = 1$, $w_{li} \in \{0,1\}$, and $Q_l = \{q_{l1}, q_{l2}, \ldots, q_{lm}\}$ is the cluster center. The algorithm adopted a simple method, called overlap measure (OM) [19], to measure the distance, as shown in Equations (2) and (3). The differences between symbols are just equal or unequal (equal is 1, unequal is 0), as shown in Equation (3).

$$d(X, Y) = \sum_{i=1}^{D} S(x_i, y_i) \qquad (2)$$

where,

$$S(x_i, y_i) = \begin{cases} 1 & if\, x_i = y_i \\ 0 & if\, x_i \neq y_i. \end{cases} \quad (3)$$

This measure method is easy to use and has great computational efficiency, since there are no involved parameters. However, its defined distances are not always reasonable in indicating the real dissimilarity because it ignores the valuable information about the relationship of the correlated attributes. There are some variants of k-mode algorithms, such as presented in [47,48]. All of these algorithms suppose that features are equally important for clustering analysis but have seen limited use in real-world practice.

In weighted features clustering algorithms, such as WKM [22], wk-modes [21], and SCC [24], features are weighted according to their importance to the clustering tasks. In these algorithms, the features are of different importance. They calculate the similarity between the two samples by supposing each dimension independently. The mathematical optimization model of these algorithms can be expressed as follows:

$$\min J(W, Q) = \sum_{l=1}^{k} \sum_{i=1}^{n} \sum_{j=1}^{m} w_{li} \lambda_{lj}^{\beta} d(X_i, Q_l) \quad (4)$$

where W is also a partition matrix and $\sum_{l=1}^{k} w_{li} = 1$, $w_{li} \in \{0,1\}$, $\Lambda = [\lambda_{lj}]$ is a weight matrix, and β is an excitation parameter which is used to control the feature weight.

The algorithm also utilized the OM method to measure the distance, as Equations (2) and (3). These methods have the advantage of high clustering efficiency. In addition, feature weighting clustering algorithms assign uniform weight to all the intra-attribute distances measured on the feature, which is suitable for well-defined distances. However, the distance measure is not well-defined for categorical data, as evidenced by the OM distance measurement. To solve this problem, most existing methods focus on exploring appropriate distance measures and attribute weighted mechanisms, such as MWKM [23]. These methods are all linear algorithms, in that they are based on the assumption that features are independent of each other, so that the relationship between features is ignored, which means that a great deal of information between the features is lost.

At present, two methods are mainly used to explore the non-linear relationship between attributes: deep neural network (DNN) and the kernel method. As we all know, DNNs need a large amount of data to train. The larger the amount of data, the more accurate the result. The kernel method uses the Mercer kernel function to implicitly describe the non-linear relationship between attributes and has been widely studied and applied because of its simplicity of mathematical expression and the high efficiency of calculation. Chen et al. [24] proposed a soft subspace clustering approach based on probabilistic distance. Its mathematical optimization model can be expressed as follows:

$$\min OBJ(\Pi, W) = \sum_{k=1}^{K} \sum_{x \in \pi_k} \sum_{d=1}^{D} w_{kd}^{\theta} Dis_d(x, \pi_k) \quad (5)$$

where W is the weight of the dth dimension for cluster k, x is the data sample and π_k is the kth cluster. $Dis_d(x, \pi_k)$ denotes the distance of sample x to the kth cluster on the dth dimension, which is computed by two discrete probabilities. This method also proposes to define a kernel density function $\kappa(X_d \mid o_{dl}; \lambda_k)$, as shown in Equation (6), to estimate the probability, where $\lambda_k \in [0,1]$ is the bandwidth for every cluster.

$$\kappa(X_d \mid o_{dl}; \lambda_k) = \begin{cases} 1 - \frac{|O_d|-1}{|O_d|} \lambda_k & X_d = o_{dl} \\ \frac{1}{|O_d|} \lambda_k & X_d \neq o_{dl} \end{cases} \quad (6)$$

where $|O_d|$ represents the power of O_d, which is the number of aggregates, and o_{dl} denotes the lth category in O_d, $o_{dl} \in O_d$.

Although this method considers the relationship between attributes in non-linear space, it does not distinguish the importance of attributes. This method also can be seen as one in which all attributes are independent of each other and all attributes in the same cluster use the same bandwidth.

3. KDE-Based Similarity for Categorical Data

In this section, we first propose a kernel density estimation (KDE) method for categorical attributes, by which each attribute has its own bandwidth. Then, the distance between categorical data objects can be expressed by a probabilistic data distribution. Moreover, a new similarity measure in the kernel subspace is defined to clustering.

3.1. Self-Expressive Kernel Density Estimation (SKDE)

Kernel density estimation method does not use the prior knowledge of the data distribution and does not attach any assumptions to data distribution. It is used to study the characteristics of data distribution from the data sample itself and is a non-parametric probability density estimation method. Unlike the kernel function seen in Equation (6), we define the kernel density function as follows:

$$\ell(X_d \mid o_{dl}; \lambda_d) = \begin{cases} 1 - \frac{|O_d|-1}{|O_d|}\lambda_d & X_d = o_{dl} \\ \frac{1}{|O_d|}\lambda_d & X_d \neq o_{dl} \end{cases} \quad (7)$$

where $\mid O_d \mid$ represents the power of O_d, which is the number of aggregates, and λ_d represents the width of the dth attribute.

It can be simply expressed as follows:

$$\ell(X_d \mid o_{dl}; \lambda_d) = \frac{1}{\mid O_d \mid}\lambda_d + (1 - \lambda_d)I(X_d = o_{dl}) \quad (8)$$

where, $I(\cdot)$ denotes the indicator function; $I(true) = 1$ and $I(false) = 0$.

According to the Equation (7), we can obtain:

$$\sum_{o_{dl} \in O_d} \ell(X_d \mid o_{dl}; \lambda_d) = 1 - \frac{\mid O_d \mid - 1}{\mid O_d \mid}\lambda_d + (\mid O_d \mid -1)\frac{\lambda_d}{\mid O_d \mid} = 1.$$

The above equation shows that the kernel function we defined satisfies the basic properties of probability distribution.

We use $\hat{p}(o_{dl} \mid \lambda_d)$ to express the kernel probability estimation of $p(o_{dl})$. According to the basic principle of the SKDE method, we have:

$$\begin{aligned} \hat{p}(o_{dl} \mid \lambda_d) &= \frac{1}{N} \sum_{x \in DB} \ell(X_d \mid o_{dl}; \lambda_d) \\ &= f(o_{dl})\left(1 - \frac{\mid O_d \mid -1}{\mid O_d \mid}\lambda_d\right) + (1 - f(o_{dl}))\frac{\lambda_d}{\mid O_d \mid} \\ &= \frac{\lambda_d}{\mid O_d \mid} + (1 - \lambda_d)f(o_{dl}) \end{aligned} \quad (9)$$

where DB is a sample set, $f(o_{dl})$ is the frequency estimation of o_{dl}.

In order to map categorical data to the high-dimensional space through the kernel function, a symbolic vectorization technique is used, as Definition 1 follows.

Definition 1. We define a data object x_{id} as follows:

$$x_{id} = \left\langle x_{id}^{(1)}, \dots, x_{id}^{(l)}, \dots, x_{id}^{(\mid O_d \mid)} \right\rangle \quad (10)$$

where $x_{id}^{(l)}$ denotes the probability of $o_{dl} \in O_d$ with regard to x_{id}, denoted by: $x_{id}^{(l)} = P_d(o_{dl}|x_d)$, and satisfies the constraint condition: $\sum_{l=1}^{|O_d|} x_{id}^{(l)} = 1$.

$x_{id}^{(l)}$ can be estimated using the kernel function shown in Equation (8), as follows:

$$\begin{aligned} x_{id}^{(l)} &= P_d(o_{dl}|x_d) \\ &\stackrel{def}{=} \ell(o_{dl}|x_d; \lambda_d) \\ &= \frac{1}{|O_d|}\lambda_d + (1 - \lambda_d)I(X_d = o_{dl}). \end{aligned} \quad (11)$$

3.2. Similarity Measurement Based on Kernel Subspace

The existing mainstream methods fail to consider the relationship between features. We formally define the non-linear similarity measurement in the kernel subspace as follows:

Definition 2. *The similarity measure of kernel subspace is given by:*

$$sim(x_i, x_j) = \kappa_w(x_i, x_j) \quad (12)$$

where $\kappa_w(x_i, x_j)$ represents the weighted features' kernel function, denoting the combination of two sample objects on each attribute.

According to Definition 2, the polynomial kernel function can be expressed as:
- origin polynomial kernel function:

$$\kappa_w(x_i, x_j) = (x_i \cdot x_j + 1)^p = \left(\sum_d^D x_{id} x_{jd} + 1\right)^p,$$

- weighted feature polynomial kernel function:

$$\kappa_w(x_i, x_j) = (x_i \cdot x_j + 1)^p = \left(\sum_d^D w_{kd}^\theta x_{id} x_{jd} + 1\right)^p.$$

We introduce a kernel function that originally acts on continuous data to project categorical data into the kernel space and a weight vector $w_k = \{w_{kd}|d = 1, 2, \ldots, D\}$ for each cluster in the kernel space for original feature selection. The greater the dth dimension's contribution to cluster, the more important it is. w_{kd} meets the constraints:

$$\begin{cases} \forall k, d : & w_{kd} \geq 0 \\ \forall k : & \sum_{d=1}^D w_{kd} = 1. \end{cases} \quad (13)$$

We introduce an index $\theta(\theta \neq 0)$ for w_{kd} to control the incentive intensity, and suppose θ as a known constant. The bigger the value of θ, the smoother the weight distribution.

This similarity measure not only uses the kernel method to "kernel" the categorical data, but also considers the relationship between features in the non-linear space. We also select features in the mapped kernel space, which distinguishes the importance of features to the cluster.

4. Proposed Clustering Algorithm

In cluster analysis, the cluster is defined as the sample set with the minimum compactness (or dispersion), in which the compactness is measured by the similarity between the sample and the cluster center. Combined with the defined non-linear similarity measurement formula of kernel subspace, the kernel subspace clustering optimization objective function of categorical data can be defined as follows:

$$J(\Pi, W) = \sum_{k=1}^{K} \sum_{x_i \in \pi_k} Sim(x_i, v_k) = \sum_{k=1}^{K} \sum_{x_i \in \pi_k} \kappa_w(x_i, v_k) \qquad (14)$$

where, v_k is the center of the cluster π_k, denoted as a D dimension vector $v_k = (v_{k1}, \ldots, v_{kd}, \ldots, v_{kD})$. Since a categorical attribute value is represented by a vector by Definition 1, so the dth dimension's center of the cluster π_k should also be represented by a vector. Each component v_{kd} represents the dth dimension's center, denoted as $v_{kd} = <v_{kd}^{(1)}, \ldots, v_{kd}^{(l)}, \ldots, v_{kd}^{(|O_d|)}>$, which meets the constraints $\sum_{l=1}^{|O_d|} v_{kd}^{(l)} = 1$, and $v_{kd}^{(l)}$ represents the probability of $o_{dl} \in O_d$ in the dth dimension.

Therefore, we have:

$$\begin{aligned} v_{kd}^{(l)} &= \frac{1}{|\pi_k|} \sum_{x_i \in \pi_k} \ell(o_{dl} | x_d; \lambda_d) \\ &= \frac{1}{|O_d|} \lambda_d + (1 - \lambda_d) f_k(o_{dl}) \end{aligned} \qquad (15)$$

where $f_k(o_{dl})$ denotes the frequency estimation of $o_{dl} \in O_d$ in the dth attribute.

4.1. Non-Linear Optimization in Kernel Subspace

In the process of calculation, the sum function is operated in the kernel function (such as the polynomial kernel subspace function mentioned above), which makes it difficult to solve w_{kd}, which, in turn, greatly increases the difficulty of solving the objective function. Therefore, we propose an efficient optimization method for solving the kernel subspace clustering optimization objective function. The objective function is transformed into the form of the existing mainstream methods (such as WKM [22] method) in order to improve the computational efficiency. The optimization objective defined by Equation (14) is further analyzed. Theorem 1 shows that for all convex kernel functions, the maximum value of Equation (14) is equivalent to the maximum value of the function of Equation (16), given by:

$$J(\Pi, W) = \sum_{k=1}^{K} \sum_{x_i \in \pi_k} \sum_{d=1}^{D} w_{kd}^{\theta} \kappa_d(x_i, v_k) \qquad (16)$$

where $\kappa_d(x_i, v_k)$ represents the mapping function's inner product of x_i and v_k in the dth dimension, that is, the kernel function in the dth dimension. For example, the polynomial kernel function can be expressed as follows:

$$\kappa_d(x_i, v_k) = (x_{id} v_{kd} + 1)^p. \qquad (17)$$

Theorem 1. *When $\theta \geq 1$, for all convex kernel functions $\kappa(\cdot, \cdot)$, the maximum Equation (14) has the same solution as the maximum Equation (16).*

Proof. We define z_d as the two input objects' combination in the dth dimension for similarity measurement in the kernel subspace. When the two input objects are the sample x_i and the cluster center v_k, z_d represents the combination of x_i and v_k in the dth dimension. If we let

$$f(\sum_{d=1}^{D} w_{kd}^{\theta} z_d) = \kappa_d(x_i, v_k),$$

in which $f(\cdot)$ is the newly defined function, we can obtain $f(z_d) = \kappa_d(x_i, v_k)$. We use mathematical induction to prove

$$\sum_{d=1}^{D} w_{kd}^{\theta} f(z_d) \leq f(\sum_{d=1}^{D} w_{kd}^{\theta} z_d).$$

(1) When $D = 1, 2$, the inequality clearly holds;

(2) We suppose that the inequality clearly holds when $D = n$, then,

$$\sum_{d=1}^{n} w_{kd}^{\theta} f(z_d) \leq f(\sum_{d=1}^{n} w_{kd}^{\theta} z_d).$$

When $D = n+1$, let $p_n = \sum_{d=1}^{n} w_{kd}$, then, we have:

$$\sum_{d=1}^{n+1} w_{kd}^{\theta} f(z_d) = w_{k(n+1)}^{\theta} f(z_{n+1}) + \sum_{d=1}^{n} w_{kd}^{\theta} f(z_d)$$

$$= w_{k(n+1)}^{\theta} f(z_{n+1}) + p_n^{\theta} \sum_{d=1}^{n} \left(\frac{w_{kd}}{p_n}\right)^{\theta} f(z_d)$$

$$\leq w_{k(n+1)}^{\theta} f(z_{n+1}) + p_n^{\theta} f\left(\sum_{d=1}^{n} \left(\frac{w_{kd}}{p_n}\right)^{\theta} z_d\right)$$

$$\leq f\left(w_{k(n+1)}^{\theta} z_{n+1} + p_n^{\theta} \sum_{d=1}^{n} \left(\frac{w_{kd}}{p_n}\right)^{\theta} z_d\right)$$

$$= f\left(w_{k(n+1)}^{\theta} z_{n+1} + \sum_{d=1}^{n} w_{kd}^{\theta} z_d\right)$$

$$= f\left(\sum_{d=1}^{n+1} w_{kd}^{\theta} z_d\right).$$

□

We can thus obtain

$$\sum_{d=1}^{D} w_{kd}^{\theta} f(z_d) \leq f(\sum_{d=1}^{D} w_{kd}^{\theta} z_d).$$

In particular, when $\theta = 1$, the inequality is Jesson inequality. We acquire $f(\sum_{d=1}^{D} w_{kd}^{\theta} z_d)$ by stretching the lower bound $\sum_{d=1}^{D} w_{kd}^{\theta} f(z_d)$ to upper bound. Then, we adjust w_{kd} to maximize $\sum_{d=1}^{D} w_{kd}^{\theta} f(z_d)$. Through step-by-step iteration, we finally obtain the maximum of $f(\sum_{d=1}^{D} w_{kd}^{\theta} z_d)$.

Combining Definition 1 and Theorem 1, the Gaussian kernel function [49] can be expressed as follows:

$$\kappa_w(x_i, x_j) = \exp\left(-\sum_{d=1}^{D} w_{kd}^{\theta} \frac{(x_{id} - x_{jd})^2}{2\sigma^2}\right) \quad (18)$$

$$= f\left(\sum_{d=1}^{D} w_{kd}^{\theta} z_d\right)$$

where $z_d = -\frac{\|x_{id} - x_{jd}\|^2}{2\sigma^2}$, $\|\cdot\|$ is the Euclidean norm, σ^2 is variance, and $f(x) = \exp(x)$.

4.2. SKSCC Clustering Algorithm

The Gaussian kernel function is the most widely used kernel function, because it has a better performance for large, as well as small samples and has fewer parameters than other kernel functions. This paper proposes the SKSCC that takes the Gaussian kernel function to be the objective function, as shown in Equation (16). We can now transfer the Equation (16) to Equation (19), as follows:

$$\begin{cases} J(\Pi, W) = \sum_{k=1}^{K} \sum_{x_i \in \pi_k} \sum_{d=1}^{D} w_{kd}^{\theta} f(z_d) \\ f(z_d) = \exp(z_d) \\ z_d = -\dfrac{\sum_{l \in |O_d|}\left[I(x_{id}=o_{dl}) - \frac{\lambda_d}{|O_d|} - (1-\lambda_d)f_k(o_{dl})\right]^2}{2\sigma^2} \end{cases} \quad (19)$$

where σ^2 is defined as the global variance, and

$$\sigma^2 = \frac{1}{ND}\sum_{i=1}^{N}\sum_{d=1}^{D}\sum_{o \in O_d}[I(x_{id}=o)-f_k(o)]^2,$$

in which N is the number of sample set, and D is the dimension of the attributes.

Equation (19) is a non-linear optimization problem with constraints. Using Lagrange multipliers, the objective function can be transferred to Equation (20) as follows:

$$\begin{cases} \max J(\Pi, W) = \sum_{k=1}^{K} \sum_{x_i \in \pi_k} \sum_{d=1}^{D} w_{kd}^{\theta} f(z_d) + \sum_{k=1}^{K} \xi_k \left(1 - \sum_{d=1}^{D} w_{kd}\right) \\ f(z_d) = \exp(z_d) \\ z_d = -\dfrac{\sum_{l \in |O_d|}\left[I(x_{id}=o_{dl}) - \frac{\lambda_d}{|O_d|} - (1-\lambda_d)f_k(o_{dl})\right]^2}{2\sigma^2}. \end{cases} \quad (20)$$

In this paper, we use the EM algorithm to optimize $\max J(\Pi, W)$, In other words, the local optimal value of J can be obtained by the iterative method. According to this principle, we first set $\Pi = \hat{\Pi}$ to maximize $J(\hat{\Pi}, W)$, and then obtain the value W, recorded as $\hat{W}$. Next, we set $W = \hat{W}$ and then maximize $J(\Pi, \hat{W})$ to calculate Π, recorded as $\hat{\Pi}$. The two steps are calculation of $\hat{W}$ and clustering, which are detailed as follows:

(1) Weight Computing

We define K independent suboptimal objective functions, as follows:

$$\begin{cases} J_k(w_k, \lambda_k) = \sum_{x_i \in \pi_k} \sum_{d=1}^{D} w_{kd}^{\theta} f(z_d) + \xi_k \left(1 - \sum_{d=1}^{D} w_{kd}\right) \\ f(z_d) = \exp(z_d) \\ z_d = -\dfrac{\sum_{l \in |O_d|}\left[I(x_{id}=o_{dl}) - \frac{\lambda_d}{|O_d|} - (1-\lambda_d)f_k(o_{dl})\right]^2}{2\sigma^2}. \end{cases} \quad (21)$$

Let $\dfrac{\partial J_k}{\partial w_{kd}} = 0$, then:

$$\frac{\partial J_k}{\partial w_{kd}} = \theta w_{kd}^{\theta-1} \sum_{x_i \in \pi_k} f(z_d) - \xi_k = 0. \quad (22)$$

Let $\dfrac{\partial J_k}{\partial \xi_k} = 0$, then:

$$\frac{\partial J_k}{\partial \xi_k} = 1 - \sum_{d=1}^{D} w_{kd} = 0. \quad (23)$$

From Equations (22) and (23), we can obtain the representation of w_{kd} as follows:

$$w_{kd} = \frac{\left(\sum_{x_i \in \pi_k} \exp\left(-\dfrac{\sum_{l \in |O_d|}\left[I(x_{id}=o_{dl}) - \frac{\lambda_d}{|O_d|} - (1-\lambda_d)f_k(o_{dl})\right]^2}{2\sigma^2}\right)\right)^{\frac{1}{1-\theta}}}{\sum_{d=1}^{D}\left(\sum_{x_i \in \pi_k} \exp\left(-\dfrac{\sum_{l \in |O_d|}\left[I(x_{id}=o_{dl}) - \frac{\lambda_d}{|O_d|} - (1-\lambda_d)f_k(o_{dl})\right]^2}{2\sigma^2}\right)\right)^{\frac{1}{1-\theta}}}. \quad (24)$$

(2) Clustering

Cluster can be generated by dividing x_i into the cluster with the most similarity. The algorithm can be expressed as follows:

$$\begin{cases} k = \arg\max_{\forall k} \kappa_w(x_i, v_k) = \arg\max_{\forall k} \left(\exp\left(-\sum_{d=1}^{D} w_{kd}^{\theta} z_d \right) \right) \\ z_d = -\frac{\sum_{l \in |O_{dl}|} \left[I(x_{id}=o_{dl}) - \frac{\lambda_d}{|O_d|} - (1-\lambda_d) f_k(o_{dl}) \right]^2}{2\sigma^2} . \end{cases} \quad (25)$$

In summary, the algorithm is outlined in Algorithm 1. According to the algorithmic structure, SKSCC can be viewed as an extension to the k-modes clustering algorithm, by adding step (3) to update the cluster and step (5) to compute the attribute weights, both of which are proportional to the kernel bandwidth that can be learned by the objects themselves. Therefore, as the k-modes algorithm, the SKSCC algorithm can also converge in a finite number of iterations. The time complexity of SKSCC is $O(KND)$.

Algorithm 1 SKSCC clustering algorithm.

Input:

The categorical dataset DB, the number of clusters K, incentive intensity θ;

Output:

Cluster Π and weight set W.

1: Initialization:

iterations' times t, t = 0;

Set all W to $\frac{1}{D}$, that's $W(0) = \frac{1}{D}$;

Calculate bandwidth λ_d; $d = 1, 2, \cdots, D$;

Calculate global variance σ^2;

Randomly select k objects as the initial cluster center, generating initial datasets, denoted as $\Pi^{(0)}$;

2: **repeat**

3:     let $\hat{W} = W^{(t)}$, divide all the samples into clusters using Equation (25), and then get $\Pi^{(t+1)}$;

4:     Update cluster center: v_{kd};

5:     Update W: set $\hat{\Pi} = \Pi^{(t+1)}$, update weight W using Equation (24), then get $W^{(t+1)}$;

6:     $t = t+1$;

7: **until** The clustering set does not change, that is, $\Pi^{(t)} = \Pi^{(t+1)}$.

8: **return** $\Pi^{(t)}$ and $W^{(t)}$.

4.3. Optimization of Kernel Bandwidths

In light of the weight calculation formula Equation (24), the weights depend on the kernel bandwidths, which is the bandwidth optimization problem in the defined SKDE method. Here, we use the mean integrated squared error (MSE) method, which is a data-driven method for estimating optimal bandwidth. For the dth attribute, the kernel probability estimation's MSE for $o_{dl} \in O_d$ can be expressed as follows:

$$MSE(o_{dl}, \lambda_d) = E\left[\sum_{o_{dl} \in O_d} (\hat{p}(o_{dl}|\lambda_d) - p(o_{dl}))^2 \right]. \quad (26)$$

According to the definition of kernel function and the properties of expectation, the bandwidth λ_d can be obtained. The objective function of the optimal estimation of bandwidth is as follows:

$$\ell(\lambda_d) = \sum_{o_{dl} \in O_d} E\left[\left(\frac{\lambda_d}{|O_d|} + (1-\lambda_d)f(o_{dl}) - p(o_{dl})\right)^2\right]$$

$$= \sum_{o_{dl} \in O_d} (1-\lambda_d)^2 E\left[f^2(o_{dl})\right] + \qquad (27)$$

$$2\left[\frac{\lambda_d(1-\lambda_d)}{|O_d|} + (\lambda_d - 1)p(o_{dl})\right]E[f(o_{dl})] +$$

$$p^2(o_{dl}) - \frac{2\lambda_d}{|O_d|}p(o_{dl}) + \frac{\lambda_d^2}{|O_d|^2}.$$

Because of

$$f(o_{dl}) = \frac{1}{N} \sum_{x_i \in DB} I(x_{id} = o_{dl}) \qquad (28)$$

where N represents the number of samples.

Then, we have:

$$E[f(o_{dl})] = E[I(X_d = o_{dl})] = p(o_{dl}). \qquad (29)$$

Due to $Var[X] = E[X^2] - (E[X])^2, [I(\cdot)]^2 = I(\cdot)$; then, we have:

$$Var[f(o_{dl})] = \frac{1}{N} Var[I(x_{id} = o_{dl})] = \frac{1}{N}\left[p(o_{dl}) - p^2(o_{dl})\right].$$

Therefore, we obtain:

$$\ell(\lambda_d) = \left(1 - \frac{1}{|O_d|}\right)\lambda_d^2 + \left(\frac{(1-\lambda_d)^2}{N} - \lambda_d^2\right)\sigma_d^2$$

where $\sigma_d^2 = 1 - \sum_{o_{dl} \in O_d} p^2(o_{dl})$.

Let $\frac{\partial \ell(\lambda_d)}{\partial \lambda_d} = 0$, then:

$$\frac{\partial \ell(\lambda_d)}{\partial \lambda_d} = \left(1 - \frac{1}{|O_d|}\right)2\lambda_d + \left(\frac{2(1-\lambda_d)(-1)}{N} - \lambda_d^2\right)\sigma_d^2 = 0.$$

Therefore, we have:

$$\lambda_d = \frac{|O_d|\sigma_d^2}{|O_d|(N + \sigma_d^2 - N\sigma_d^2) - N}. \qquad (30)$$

We use the frequency distribution of the training samples to estimate $p(o_{dl})$, and we calculate σ_d^2 by the standard deviation of the training samples. Hence, we obtain

$$s_d^2 = 1 - \sum_{o_{dl} \in O_d} f^2(o_{dl}). \qquad (31)$$

The kernel bandwidth algorithm is outlined in Algorithm 2. Several properties of the kernel bandwidth's optimal estimation are analyzed:

(1) The larger the number of samples N, the smaller the bandwidth.

$$\lambda_d^* = \frac{|O_d|s_d^2}{|O_d|(N + \sigma_d^2 - N\sigma_d^2) - N}$$

$$= \frac{s_d^2}{N\left(\sum_{o_{dl} \in O_d} f^2(o_{dl}) - \frac{1}{|O_d|}\right) + s_d^2}$$

The coefficient of N is $\sum_{o_{dl} \in O_d} f^2(o_{dl}) - \frac{1}{|O_d|}$; its values' range is $[0, 1]$. The larger the number of samples N, the smaller the bandwidths. When $N \to \infty$, the bandwidth

$\lambda_d \to 0$. This is consistent with the effect of bandwidth as the smoothing parameter of the kernel function.

(2) The larger the data dispersion, the larger the bandwidth.

$$\lambda_d^* = \frac{|O_d|s_d^2}{|O_d|(N+\sigma_d^2 - N\sigma_d^2) - N}$$

$$= \frac{s_d^2}{N - \frac{N}{|O_d|} - (N-1)s_d^2}$$

Let us calculate the derivative of λ_d^* with respect to s_d^2 as follows:

$$\frac{\partial \lambda_d^*}{\partial s_d^2} = \frac{N\left(1 - \frac{1}{|O_d|}\right)}{\left(N - \frac{N}{|O_d|} - (N-1)s_d^2\right)^2}.$$

Because $1 - \frac{1}{|O_d|} > 0$, then $\frac{\partial \lambda_d^*}{\partial s_d^2} > 0$; so, λ_d^* is the increasing function with respect to s_d^2 in the range [0,1). The larger the data dispersion s_d^2, the larger the bandwidth λ_d^*, that is to say, the larger the discreteness of an attribute, the larger the kernel bandwidth corresponding to the attribute. In particular, when an attribute categorical data are uniformly distributed, the corresponding kernel bandwidth takes the maximum value.

Algorithm 2 The kernel bandwidth calculation algorithm.

Input:
  The categorical dataset DB;
Output:
  $\Lambda = \{\lambda_d | d = 1, 2, \ldots, D\}$;
1: **for** $d = 1$ to D **do**
2:   Compute s_d^2 using Equation (23);
3:   Compute λ_d using Equation (22);
4: **end for**

---

5. Experimental Analysis

Experiments were performed to verify the effectiveness of our proposed SKSCC on synthetic and real datasets. Comparative experiments were carried out on some current mainstream categorical clustering algorithms.

5.1. Experimental Setup

In practical applications, the Gaussian kernel function is the most widely used kernel function, because it is suitable for a variety of samples and has few parameters. Moreover, the mapping space provided by this type of kernel function isinfinitely dimensional, so that the data that are not separated in the original space can be directly mapped into linear separable points. Therefore, we chose the Gaussian kernel to mine the non-linear relationship between categorical attributes. The parameter defined as

$$\sigma^2 = \frac{1}{ND} \sum_{i=1}^{N} \sum_{d=1}^{D} \sum_{o \in O_d} \left(I(x_{id} = o) - \frac{\lambda_d}{|O_d|} - (1-\lambda_d)f(o_{dl}) \right)^2 \qquad (32)$$

which is the global variance, and is learned from the data themselves.

We chose three algorithms—k-mode [17], WKM [22], MWKM [23]—for our comparative experiments. WKM introduced attributes-weighting within the framework of the k-modes algorithm, which is a linear weighting. The MWKM algorithm weights the attributes through the frequency of the mode. All three methods are based on the principle of feature independence to calculate the sample similarity (or dissimilarity). These algorithms are selected for comparison with the non-linear similarity measurement SKSCC. The parameter β is set to 2 in WKM. The parameter β is set to 2 and $T_s = T_v = 1$ in MWKM.

Synthetic data can control the cluster structure of datasets through the number and size of clusters, which is conducive for analyzing the performance of the algorithm and its adaptability to various datasets. For this paper, we first tested on several synthetic datasets and then carried out experiments on many real datasets. Because the labels are all known, two external evaluation indices—accuracy and F-score [22]—were selected to evaluate the clustering performance of the new algorithm. The larger the value of the two indices, the better the clustering effect. F-score is defined as follows:

$$F-score = \sum_{k=1}^{K} \frac{n_k}{N} \max_{1 \leq i \leq K} \left[\frac{2 \times R(class_k, \pi_i) \times P(class_k, \pi_i)}{R(class_k, \pi_i) + P(class_k, \pi_i)} \right]$$

where $class_k$ represents the kth real class in datasets, n_k represents the sample number of $class_k$, and $P(class_k, \pi_i)$ and $R(class_k, \pi_i)$ separately represent accuracy and recall compared real class $class_k$ and cluster π_i of clustering results, that is,

$$P = \frac{TP}{TP + FP}$$

$$R = \frac{TP}{TP + FN}$$

where TP represents the number of predicting correct clusters as correct clusters; FN represents the number of predicting correct clusters as false clusters; FP represents the number of predicting false clusters as correct clusters.

5.2. Discussion of Parameters

In the kernel space, each attribute is automatically given a weight to measure its similarity, and the corresponding subspace is found through feature selection.

$$w_{kd}^{\theta} = \frac{\left(\sum_{x_i \in \pi_k} \exp\left(-\frac{\sum_{l \in |O_d|} \left[I(x_{id} = o_{dl}) - \frac{\lambda_d}{|O_d|} - (1-\lambda_d) f_k(o_{dl}) \right]^2}{2\sigma^2} \right) \right)^{\frac{\theta}{1-\theta}}}{\sum_{d=1}^{D} \left(\sum_{x_i \in \pi_k} \exp\left(-\frac{\sum_{l \in |O_d|} \left[I(x_{id} = o_{dl}) - \frac{\lambda_d}{|O_d|} - (1-\lambda_d) f_k(o_{dl}) \right]^2}{2\sigma^2} \right) \right)^{\frac{\theta}{1-\theta}}}$$

where θ is the incentive intensity, and is the allocation parameter of control weight. Figure 1 shows the change in parameters for the weight of the three attributes in the Breastcancer dataset. Here, the discreteness of the three attributes is set to increase from attribute 1. There are four comments for θ.

(1) When $\theta = 0$, w_{kd}^{θ} is the constant, that is, each attribute will be assigned an equal weight;
(2) When $\theta = 1$, $\frac{\theta}{1-\theta} \to \infty$, but all of the weights must meet the restriction $\sum_{d=1}^{D} w_{kd} = 1$, so when $\theta \to 1^+$, the attribute with the minimum deviation of the sample will be weighted, while the rest of the attributes will be given zero weight; when $\theta \to 1^-$, the importance of all attributes tends to be the same;
(3) When $0 < \theta < 1$, the more discrete the attribute, the greater its weight;
(4) When $\theta < 0$ and $\theta > 1$, the attribute weight is inversely proportional to the dispersion of data distribution. Considering Theorem 1, we should set $\theta > 1$, but when θ is too larger, the difference between attribute weights is reduced.

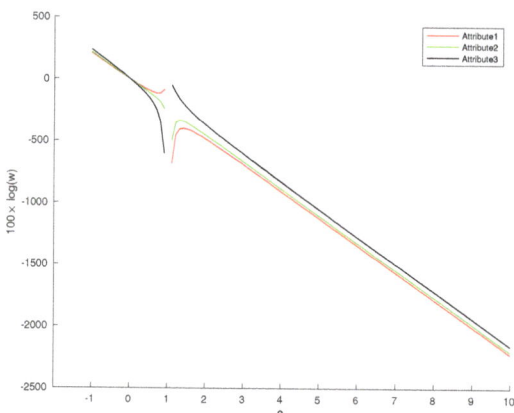

Figure 1. Analysis of weight with different θ.

5.3. Analysis of Synthetic Data and Results

This study used MATLAB (Version 9.9.0.1495850 R2020b) to generate the synthetic data in the experiment. First, four multi-dimensional numerical datasets were generated by the MATLAB function $mvnrnd(\cdot)$, in which the weight of attributes was controlled by setting the variance of attributes, and the correlation degree between attributes was controlled by adjusting the parameters of the covariance matrix. The synthesized numerical data were then discretized by equal width [40] and transformed into categorical data. The synthetic datasets that contain the correct category labels are presented in Table 1. Four datasets were used to verify the advantages of SKSCC compared with the current mainstream categorical clustering methods.

- The covariance of attribute 1 and attribute 2 is set to -2 in DataSet1, which makes their attributes a negative correlation. The covariance of attribute 1 and attribute 4 is set to 2, which makes their attributes a positive correlation. The variances are set to be equal on each attribute;
- DataSet2 and DataSet1 are set to the same clusters, but the number of attributes differs. Ten attributes are extracted to set their covariance. The variances are set to be equal on each attribute;
- DataSet3 and DataSet2 are set to the same attributes, but the clusters are different. The variances are set to be equal in two clusters. Ten attributes are extracted to set their covariance;
- DataSet4 is set to the most number of attributes and the clusters. Twenty attributes are extracted to set their covariance in seven clusters. All attributes are set to covariance in one cluster. A half clusters set the same variances, as well as other half clusters.

Table 1. Data categorized in four synthetic datasets.

	Attributes (D)	Clusters (K)	Samples (N)
Datasets1	6	2	1000
Datasets2	20	2	1000
Datasets3	20	4	1000
Datasets4	40	8	1000

We implemented 100 runs on each algorithm and each dataset, and set $\theta = 1.5$. The average clustering accuracy reported in Table 2 reflects the overall performance of each clustering algorithm, and the stability of clustering performance of each algorithm can be judged according to the listed variance. The smaller the variance of clustering accuracy, the better the stability of clustering performance.

Table 2. Comparison of F-score and Accuracy results of four algorithms performed on the four synthetic datasets.

Index	Datasets	K-Mode [17]	WKM [22]	MWKM [23]	SKSCC
F-Score	Datasets1	0.9823 ± 0.0000	0.9489 ± 0.0079	0.9738 ± 0.0018	1.0000 ± 0.0000
	Datasets2	0.9762 ± 0.0015	0.9860 ± 0.0000	0.9860 ± 0.0000	0.9940 ± 0.0000
	Datasets3	0.6346 ± 0.0011	0.5766 ± 0.0018	0.6311 ± 0.0009	0.6771 ± 0.0005
	Datasets4	0.5268 ± 0.0008	0.3839 ± 0.0033	0.5367 ± 0.0010	0.6224 ± 0.0017
Accuracy	Datasets1	0.9823 ± 0.0000	0.9589 ± 0.0038	0.9746 ± 0.0012	1.0000 ± 0.0000
	Datasets2	0.9762 ± 0.0015	0.9860 ± 0.0000	0.9860 ± 0.0000	0.9939 ± 0.0000
	Datasets3	0.6755 ± 0.0016	0.6037 ± 0.0024	0.6644 ± 0.0009	0.7033 ± 0.0004
	Datasets4	0.5863 ± 0.0013	0.5053 ± 0.0147	0.5848 ± 0.0014	0.6655 ± 0.0014

From Table 2, we can see that with the increase in the number of related attributes, the clustering accuracy of SKSCC is significantly higher than that of other algorithms. This is because SKSCC employs a "kernel" operation and take into consideration the relationship between attributes.

5.4. Analysis of Real-World Data and Results

In this part of the experiments, we set out to test and verify the performance of SKSCC in real-world datasets. We compared the SKSCC algorithm with three other algorithms: the original k-modes algorithm (k-mode), the weighting algorithm (WKM), and the mixed weighting algorithm (MWKM).

5.4.1. Real-World Datasets

To carry out the experiments, we obtained 10 datasets from the University of California Irvine (UCI) Machine Learning Repository [7]. Table 3 lists the details of these 10 datasets. The Breastcancer, Vote, Mushroom, and Adult+stretch datasets have the same clusters, but Mushroom dataset has the most samples, and Adult+stretch dataset has the least number of samples. The Balance and Splice datasets each have the same number of clusters (3), but the dimensionality of Splice is higher. The Soybeansmall and Car datasets each have the same number of clusters (4), but different attributes and samples. Dermatology and Zoo are multi-cluster datasets.

Table 3. Details of 10 DataSets from UCI.

No.	UCI Datasets	Attributes (D)	Clusters (K)	Samples (N)
1	Breastcancer	9	2	699
2	Vote	16	2	435
3	Mushroom	21	2	8124
4	Adult+stretch	4	2	20
5	Balance	4	3	625
6	Splice	60	3	3190
7	Soybeansmall	35	4	47
8	Car	6	4	1728
9	Dermatology	33	6	366
10	Zoo	15	7	101

5.4.2. Comparison of Clustering Quality

Because the initial cluster centers can affect the algorithm results, we randomly selected 100 initial centers, and all of the algorithms used the same initial centers in each experiment. We implemented 100 runs on each algorithm and each dataset, and set $\theta = 1.5$. The average values and the errors for F-score and accuracy are presented in Table 4. The results showed that our proposed method, SKSCC, achieved the best performance in the comparative experiments on most of the datasets. Because the k-mode [17], WKM [22], and MWKM [23] algorithms are all based on the mode-type category theory, it is easy for them to descend to the clustering objective algorithm's local minimum, causing them to lose applicability. However, WKM achieved good results on the Car and Splice datasets, while MWKM achieved high accuracy on the Dermatology dataset.

Table 4. Comparison of clustering results in terms of F-score and accuracy.

Index	Datasets	K-Mode [17]	WKM [22]	MWKM [23]	SKSCC
F-Score	Breastcancer	0.8637 ± 0.0000	0.7683 ± 0.0005	0.8645 ± 0.0155	0.9660 ± 0.0000
	Vote	0.8610 ± 0.0000	0.8238 ± 0.0073	0.8698 ± 0.0000	0.8749 ± 0.0000
	Mushroom	0.7159 ± 0.0171	0.6645 ± 0.0034	0.7480 ± 0.0202	0.7901 ± 0.0193
	Adult + stretch	0.6691 ± 0.0135	0.6722 ± 0.0159	0.6876 ± 0.0163	0.7537 ± 0.0085
	Balance	0.4882 ± 0.0016	0.4782 ± 0.0022	0.4630 ± 0.0024	0.5672 ± 0.0017
	Splice	0.4155 ± 0.0000	0.5321 ± 0.0007	0.4313 ± 0.0000	0.5258 ± 0.0019
	Soybeansmall	0.8324 ± 0.0152	0.7336 ± 0.0157	0.8436 ± 0.0175	0.8641 ± 0.0146
	Car	0.4412 ± 0.0018	0.5006 ± 0.0057	0.4268 ± 0.0012	0.4738 ± 0.0028
	Dermatology	0.6476 ± 0.0083	0.5573 ± 0.0136	0.6685 ± 0.0088	0.6357 ± 0.0034
	Zoo	0.7273 ± 0.0090	0.6716 ± 0.0130	0.7417 ± 0.0074	0.7701 ± 0.0070
Accuracy	Breastcancer	0.8621 ± 0.0000	0.8284 ± 0.0000	0.8659 ± 0.0156	0.9659 ± 0.0000
	Vote	0.8625 ± 0.0000	0.8244 ± 0.0066	0.8681 ± 0.0000	0.8734 ± 0.0000
	Mushroom	0.7536 ± 0.0134	0.8481 ± 0.0157	0.7733 ± 0.0143	0.8194 ± 0.0131
	Adult + stretch	0.7150 ± 0.0160	0.7165 ± 0.0168	0.6910 ± 0.0159	0.8620 ± 0.0086
	Balance	0.5251 ± 0.0010	0.4629 ± 0.0033	0.4327 ± 0.0024	0.8722 ± 0.0321
	Splice	0.4237 ± 0.0000	0.6149 ± 0.0011	0.4314 ± 0.0000	0.5426 ± 0.0017
	Soybeansmall	0.8740 ± 0.0110	0.9423 ± 0.0039	0.8915 ± 0.0110	0.9085 ± 0.0083
	Car	0.4023 ± 0.0013	0.4550 ± 0.0095	0.3593 ± 0.0000	0.4251 ± 0.0038
	Dermatology	0.7085 ± 0.0076	0.9298 ± 0.0038	0.7367 ± 0.0063	0.6911 ± 0.0048
	Zoo	0.7937 ± 0.0066	0.8260 ± 0.0084	0.7895 ± 0.0073	0.8043 ± 0.0061

Figure 2 shows the distribution of clustering accuracy for all the algorithms when run 100 times. SKSCC has the best stability. The abscissa represents each algorithm's running time, and the ordinate is the F-score value to express the results of each clustering. SKSCC has the smallest fluctuation among all the algorithms, although WKM has the best average F-score on the Splice and Car datasets, and MWKM has the best average F-score on the Dermatology dataset. The clustering results for the k-mode algorithms show significant contrast, because they consider only the module in the clustering process, which makes it easy to fall into the local optimum, and the initial cluster center is k randomly selected objects. This is reflected in the standard deviation of average precision. Because SKSCC quantizes the module, it avoids the above-mentioned problems and has more stable performance than the other algorithms.

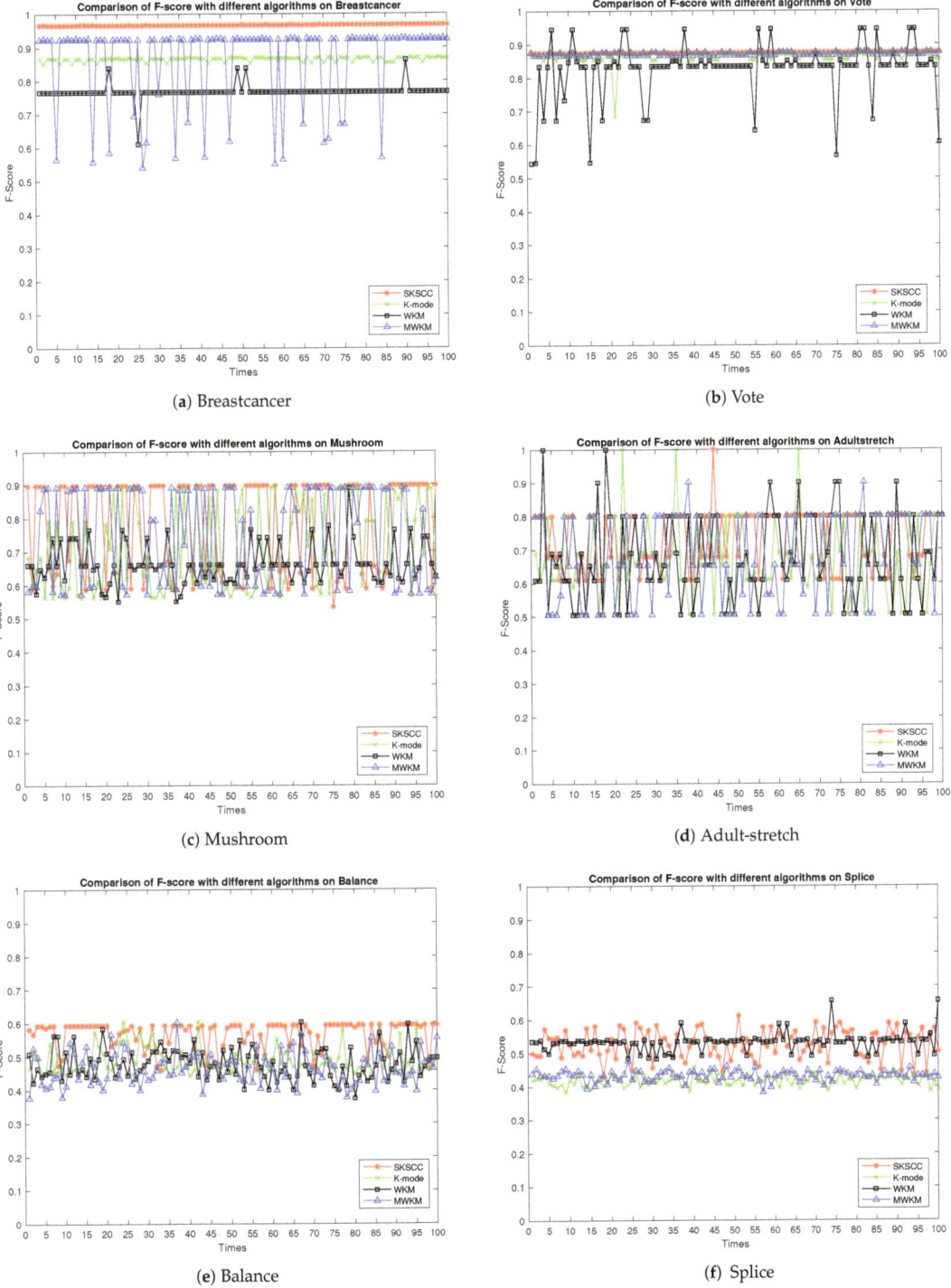

Figure 2. *Cont.*

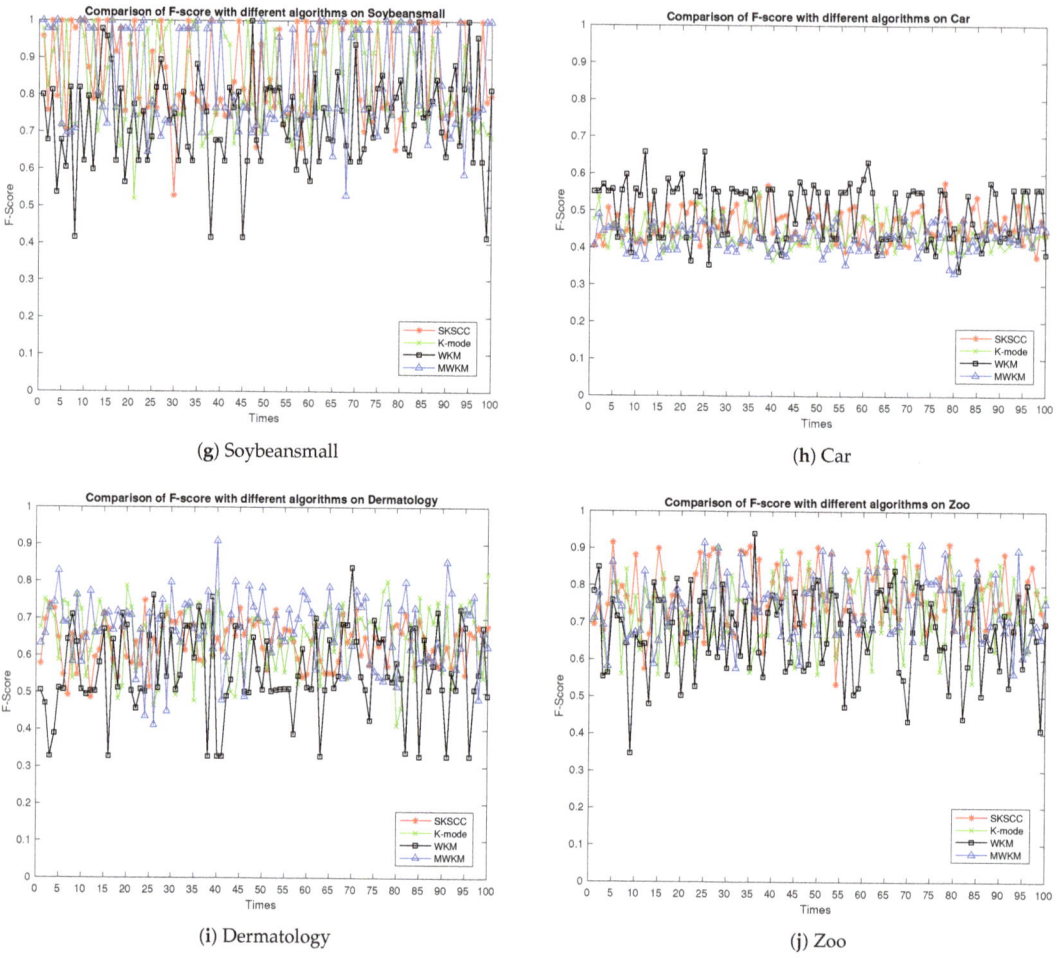

Figure 2. Comparison of F-score with different algorithms on different datasets.

5.4.3. Feature Weighting Results

Our SKSCC approach also has a feature selection effect. Using the Breastcancer dataset as an example, Figure 3 shows the attribute weights generated by the MWKM and SKSCC algorithms. It does not show the k-mode algorithm or WKM algorithm, because the former method is not weighted in its features and the latter method calculates the weights based on mode frequency, which is similar to MWKM algorithm. From Figure 4, we can see that for SKSCC, A1 and A9 acquire the largest and the smallest weights, respectively, of the benign class, but MWKM algorithm achieved the opposite results. To test the feature weighting method's rationality for the SKSCC, we removed the A1 and A9 features from the original Breastcancer data in order to form two reduced datasets. The F-score values of the different clustering algorithms on the Breastcancer dataset with the original and reduced feature sets are shown in Figure 4. For all the algorithms, the reduced dataset with the A9 feature removed achieved the highest F-score values, while the reduced dataset with the A1 feature removed showed decreased F-score values. The results indicate that our SKSCC algorithm with non-linear similarity measurement does a better job, by considering the relationship of the attributes, than the other algorithms.

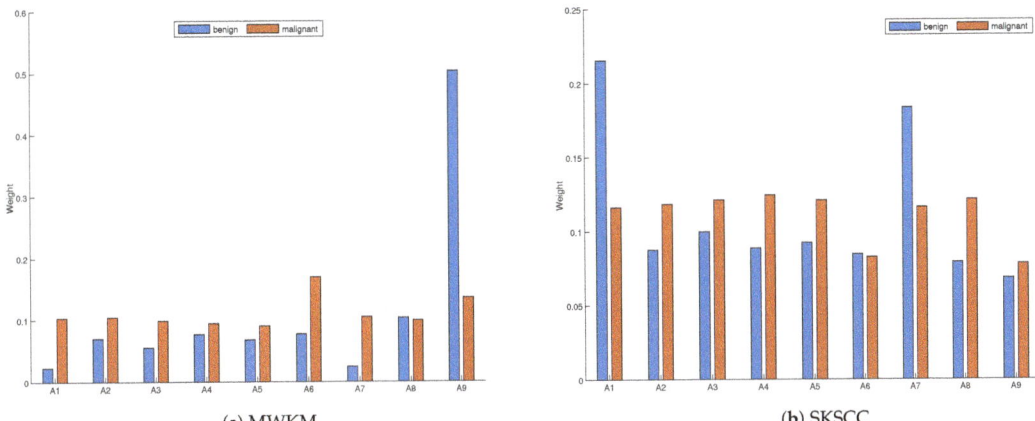

Figure 3. Weight distributions generated by two algorithms on Breastcancer dataset.

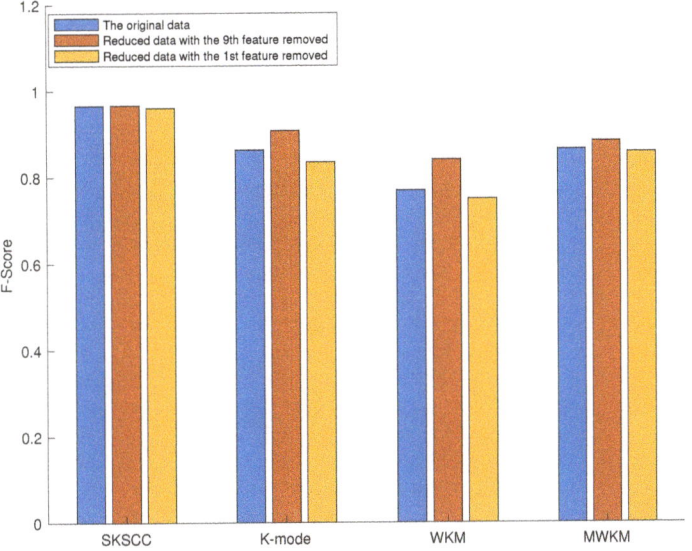

Figure 4. F-score values of the different clustering algorithms on the Breastcancer dataset with original and reduced feature sets.

5.4.4. Time Consumption

This paper uses a logarithm of the average time of clustering to compare the actual average times. The ordinate represents the average time (in MS) of each algorithm running on the real-world dataset. It can be seen from Figure 5 that k-mode, WKM, and MWKM algorithms have high clustering efficiency, which is one of the advantages of the module-based clustering algorithms. Because only the module of the categorical attribute needs to be considered, the statistical information of the other categorical symbols can be ignored, which greatly reduces the algorithms' clustering times.

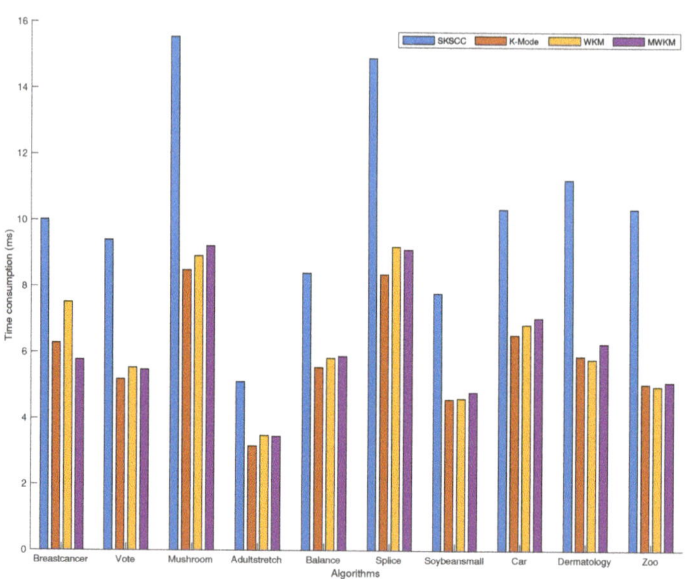

Figure 5. F-Score values of the different clustering algorithms on the Breastcancer dataset with original and reduced feature sets.

6. Conclusions

Kernel clustering with categorical data is a vital direction in application research. In view of current problems, such as supposing all features independently, considering all attributes' importance equally, and finding an optimization solution, this paper proposes a novel kernel clustering approach for categorical data, that is, a self-expressive kernel subspace clustering algorithm for categorical data (SKSCC). This paper first defines a kernel function for self-expression kernel density estimation (SKDE), in which each attribute has its own bandwidth and can be calculated by the data themselves. We also propose a novel non-linear similarity measurement method and an efficient non-linear optimization method (Theorem 1) to solve the objective function of the kernel clustering. Finally, the SKSCC algorithm is presented for categorical data. Our method not only considers the relationship between attributes in non-linear space but also gives each attribute a feature weight to measure the correlation degree in the algorithmic process. The experimental results indicate that the proposed algorithm outperforms the other algorithms on the synthetic and UCI datasets.

There are many directions that are of interest for future exploration. We will expand our approach to other kernel functions and test the performance on more datasets for various data. Our efforts will also be directed at combining our method with deep learning to estimate the parameters adaptively.

Author Contributions: Conceptualization, Q.J. and L.C.; methodology, H.C. and K.X.; software, H.C.; validation, H.C. and K.X.; formal analysis, H.C. and K.X.; investigation, Q.J. and L.C.; resources, Q.J. and L.C.; data curation, H.C.; writing—original draft preparation, H.C.; writing—review and editing, H.C.; visualization, H.C.; supervision, Q.J. and L.C.; project administration, Q.J. and L.C.; funding acquisition, Q.J. and L.C. All authors have read and agreed to the published version of the manuscript.

Funding: The work is supported by the Key-Area Research and Development Program of Guangdong Province Grant No. 2019B010137002, and the National Natural Science Foundation of China under Grant Nos. U1805263, 61672157.

Institutional Review Board Statement: Not applicable.

Informed Consent Statement: Not applicable.

Data Availability Statement: Not applicable.

Acknowledgments: The authors would like to thank all the anonymous reviewers for their insightful comments and constructive suggestions that have obviously upgraded the quality of this manuscript.

Conflicts of Interest: The authors declare that they have no known competing financial interests or personal circumstances that could have appeared to influence the work reported in this manuscript.

References

1. Tang, J.; Liu, H. An unsupervised feature selection framework for social media data. *IEEE Trans. Knowl. Data Eng.* **2014**, *26*, 2914–2927. [CrossRef]
2. Alelyani, S.; Tang, J.; Liu, H. Feature selection for clustering: A review. *Data Clust. Algorithms Appl.* **2013**, *29*, 144.
3. Han, J.; Kamber, M. *Data Mining: Concepts and Techniques*; Morgan Kaufmann: San Francisco, CA, USA, 2001.
4. Bharti, K.K.; Singh, P.K. A survey on filter techniques for feature selection in text mining. In Proceedings of the Second International Conference on Soft Computing for Problem Solving (SocProS 2012), Jaipur, India, 28–30 December 2012; Springer: New Delhi, India, 2014; pp. 1545–1559.
5. Yasmin, M.; Mohsin, S.; Sharif, M. Intelligent image retrieval techniques: A survey. *J. Appl. Res. Technol.* **2014**, *12*, 87–103. [CrossRef]
6. Saeys, Y.; Inza, I.; Larranaga, P. A review of feature selection techniques in bioinformatics. *Bioinformatics* **2007**, *23*, 2507–2517. [CrossRef]
7. Frank, A. UCI Machine Learning Repository. 2010. Available online: http://archive.ics.uci.edu/ml (accessed on 28 March 2021).
8. Jain, A.K.; Murty, M.N.; Flynn, P.J. Data clustering: A review. *ACM Comput. Surv. (CSUR)* **1999**, *31*, 264–323. [CrossRef]
9. Xu, R.; Wunsch, D. Survey of clustering algorithms. *IEEE Trans. Neural Netw.* **2005**, *16*, 645–678. [CrossRef]
10. Jain, A.K. Data clustering: 50 years beyond k-mean. *Pattern Recognit. Lett.* **2010**, *31*, 651–666. [CrossRef]
11. Wu, S.; Lin, J.; Zhang, Z.; Yang, Y. Hesitant fuzzy linguistic agglomerative hierarchical clustering algorithm and its application in judicial practice. *Mathematics* **2021**, *9*, 370. [CrossRef]
12. Guha, S.; Rastogi, R.; Shim, K. ROCK: A robust clustering algorithm for categorical attributes. *Inf. Syst.* **2000**, *25*, 345–366. [CrossRef]
13. Andritsos, P.; Tzerpos, V. Information-theoretic software clustering. *IEEE Trans. Softw. Eng.* **2005**, *31*, 150–165. [CrossRef]
14. Andritsos, P.; Tsaparas, P.; Miller, R.J.; Sevcik, K.C. LIMBO: Scalable clustering of categorical data. In Proceedings of the International Conference on Extending Database Technology, Heraklion, Crete, Greece, 14–18 March 2004; Springer: Berlin/Heidelberg, Germany, 2004; pp. 123–146.
15. Qin, H.; Ma, X.; Herawan, T.; Zain, J.M. MGR: An information theory based hierarchical divisive clustering algorithm for categorical data. *Knowl.-Based Syst.* **2014**, *67*, 401–411. [CrossRef]
16. Xiong, T.; Wang, S.; Mayers, A.; Monga, E. DHCC: Divisive hierarchical clustering of categorical data. *Data Min. Knowl. Discov.* **2012**, *24*, 103–135. [CrossRef]
17. Huang, Z. Extensions to the k-means algorithm for clustering large data sets with categorical values. *Data Min. Knowl. Discov.* **1998**, *2*, 283–304. [CrossRef]
18. Huang, Z.; Ng, M.K. A fuzzy k-modes algorithm for clustering categorical data. *IEEE Trans. Fuzzy Syst.* **1999**, *7*, 446–452. [CrossRef]
19. Ng, M.K.; Li, M.J.; Huang, J.Z.; He, Z. On the impact of dissimilarity measure in k-modes clustering algorithm. *IEEE Trans. Pattern Anal. Mach. Intell.* **2007**, *29*, 503–507. [CrossRef] [PubMed]
20. Bai, L.; Liang, J.; Dang, C.; Cao, F. The impact of cluster representatives on the convergence of the k-modes type clustering. *IEEE Trans. Pattern Anal. Mach. Intell.* **2012**, *35*, 1509–1522. [CrossRef] [PubMed]
21. Cao, F.; Liang, J.; Li, D.; Zhao, X. A weighting k-modes algorithm for subspace clustering of categorical data. *Neurocomputing* **2013**, *108*, 23–30. [CrossRef]
22. Chan, E.Y.; Ching, W.K.; Ng, M.K.; Huang, J.Z. An optimization algorithm for clustering using weighted dissimilarity measures. *Pattern Recognit.* **2004**, *37*, 943–952. [CrossRef]
23. Bai, L.; Liang, J.; Dang, C.; Cao, F. A novel attribute weighting algorithm for clustering high-dimensional categorical data. *Pattern Recognit.* **2011**, *44*, 2843–2861. [CrossRef]
24. Chen, L.; Wang, S.; Wang, K.; Zhu, J. Soft subspace clustering of categorical data with probabilistic distance. *Pattern Recognit.* **2016**, *51*, 322–332. [CrossRef]
25. Han, J.; Kamber, M.; Pei, J. Data mining concepts and techniques third edition. *Morgan Kaufmann Ser. Data Manag. Syst.* **2011**, *5*, 83–124.
26. Guyon, I.; Elisseeff, A. An introduction to variable and feature selection. *J. Mach. Learn. Res.* **2003**, *3*, 1157–1182.
27. Breiman, L.; Friedman, J.; Stone, C.J.; Olshen, R.A. *Classification and Regression Trees*; CRC Press: Boca Raton, FL, USA 1984.
28. Kohavi, R.; John, G.H. Wrappers for feature subset selection. *Artif. Intell.* **1997**, *97*, 273–324. [CrossRef]

29. Pashaei, E.; Aydin, N. Binary black hole algorithm for feature selection and classification on biological data. *Appl. Soft Comput.* **2017**, *56*, 94–106. [CrossRef]
30. Rasool, A.; Tao, R.; Kamyab, M.; Hayat, S. Gawa—A feature selection method for hybrid sentiment classification. *IEEE Access* **2020**, *8*, 191850–191861. [CrossRef]
31. Liu, H.; Setiono, R. Chi2: Feature selection and discretization of numeric attributes. In Proceedings of the 7th IEEE International Conference on Tools with Artificial Intelligence, Herndon, VA, USA, 5–8 November 1995; pp. 388–391.
32. Quinlan, J.R. Induction of decision trees. *Mach. Learn.* **1986**, *1*, 81–106. [CrossRef]
33. Quinlan, J.R. *C4. 5: Programs for Machine Learning*; Elsevier: Amsterdam, The Netherlands, 2014.
34. Kandaswamy, K.K.; Pugalenthi, G.; Hazrati, M.K.; Kalies, K.U.; Martinetz, T. BLProt: Prediction of bioluminescent proteins based on support vector machine and relieff feature selection. *BMC Bioinform.* **2011**, *12*, 345. [CrossRef] [PubMed]
35. Shao, J.; Liu, X.; He, W. Kernel based data-adaptive support vector machines for multi-class classification. *Mathematics* **2021**, *9*, 936. [CrossRef]
36. Robnik-Šikonja, M.; Kononenko, I. Theoretical and empirical analysis of ReliefF and RReliefF. *Mach. Learn.* **2003**, *53*, 23–69. [CrossRef]
37. Le, T.T.; Urbanowicz, R.J.; Moore, J.H.; McKinney, B.A. Statistical inference Relief (STIR) feature selection. *Bioinformatics* **2019**, *35*, 1358–1365. [CrossRef]
38. Huang, Z.; Yang, C.; Zhou, X.; Huang, T. A hybrid feature selection method based on binary state transition algorithm and ReliefF. *IEEE J. Biomed. Health Inform.* **2018**, *23*, 1888–1898. [CrossRef] [PubMed]
39. Deng, Z.; Chung, F.L.; Wang, S. Robust relief-feature weighting, margin maximization, and fuzzy optimization. *IEEE Trans. Fuzzy Syst.* **2010**, *18*, 726–744. [CrossRef]
40. Chen, L.F. A probabilistic framework for optimizing projected clusters with categorical attributes. *Sci. China Inf. Sci.* **2015**, *58*, 1–15. [CrossRef]
41. Kong, R.; Zhang, G.; Shi, Z.; Guo, L. Kernel-based k-means clustering. *Comput. Eng.* **2004**, *30*, 12–14.
42. Elhamifar, E.; Vidal, R. Sparse subspace clustering: Algorithm, theory, and applications. *IEEE Trans. Pattern Anal. Mach. Intell.* **2013**, *35*, 2765–2781. [CrossRef]
43. Ji, P.; Zhang, T.; Li, H.; Salzmann, M.; Reid, I. Deep subspace clustering networks. *arXiv* **2017**, arXiv:1709.02508.
44. You, C.; Li, C.G.; Robinson, D.P.; Vidal, R. Oracle based active set algorithm for scalable elastic net subspace clustering. In Proceedings of the IEEE Conference on Computer Vision and Pattern Recognition, Las Vegas, NV, USA, 26 June–1 July 2016; pp. 3928–3937.
45. Chen, L.; Guo, G.; Wang, S.; Kong, X. Kernel learning method for distance-based classification of categorical data. In Proceedings of the 2014 14th UK Workshop on Computational Intelligence (UKCI), Bradford, UK, 8–10 September 2014; pp. 1–7.
46. Ouyang, D.; Li, Q.; Racine, J. Cross-validation and the estimation of probability distributions with categorical data. *J. Nonparametr. Stat.* **2006**, *18*, 69–100. [CrossRef]
47. Huang, Z. Clustering large data sets with mixed numeric and categorical values. In Proceedings of the 1st Pacific-Asia Conference on Knowledge Discovery and Data Mining (PAKDD), Singapore, 23–24 February 1997; pp. 21–34.
48. Cheung, Y.M.; Jia, H. Categorical-and-numerical-attribute data clustering based on a unified similarity metric without knowing cluster number. *Pattern Recognit.* **2013**, *46*, 2228–2238. [CrossRef]
49. Zhong, S.; Chen, D.; Xu, Q.; Chen, T. Optimizing the gaussian kernel function with the formulated kernel target alignment criterion for two-class pattern classification. *Pattern Recognit.* **2013**, *46*, 2045–2054. [CrossRef]

MDPI
St. Alban-Anlage 66
4052 Basel
Switzerland
Tel. +41 61 683 77 34
Fax +41 61 302 89 18
www.mdpi.com

Mathematics Editorial Office
E-mail: mathematics@mdpi.com
www.mdpi.com/journal/mathematics

www.ingramcontent.com/pod-product-compliance
Lightning Source LLC
LaVergne TN
LVHW070657100526
838202LV00013B/986